Crystallographic Computing 3: Data Collection, Structure Determination, Proteins, and Databases

Papers presented at the International Summer School on Crystallographic Computing held at the Max-Planck-Institut für Kohlenforschung, Mülheim an der Ruhr FRG, 30 July–8 August 1984

Edited by

G. M. Sheldrick
Institut für Anorganische Chemie
Universität Göttingen, FRG

and

C. Krüger and R. Goddard
Max-Planck-Institut für Kohlenforschung,
Mülheim an der Ruhr, FRG

CLARENDON PRESS · OXFORD
1985

Oxford University Press, Walton Street, Oxford OX2 6DP
Oxford New York Toronto
Delhi Bombay Calcutta Madras Karachi
Kuala Lumpur Singapore Hong Kong Tokyo
Nairobi Dar es Salaam Cape Town
Melbourne Auckland
and associate companies in
Beirut Berlin Ibadan Nicosia

Published in the United States
by Oxford University Press, New York

British Library Cataloguing in Publication Data
International Summer School on Crystallographic Computing (1984: Mülheim an der Ruhr).
Data collection, structure determination, proteins and databases : papers presented at the International Summer School on Crystallographic Computing held at the Max-Planck-Institut für Kohlenforschung, Mülheim an der Ruhr FRG, 30 July–8 August 1984. – Crystallographic computing; 3)
1. Crystallography – Data processing
I. Title II. Sheldrick, G.M. III. Krüger, C.
IV. Goddard, R. V. Series
548'.028'54 QD906.7.E4
ISBN 0-19-855211-4

Library of Congress Cataloging in Publication Data
International Summer School on Crystallographic Computing (9th: 1984: Max-Planck-Institut für Kohlenforschung)
Crystallographic computing 3.
"Papers presented at the International Summer School on Crystallographic Computing held at the Max-Planck-Institut für Kohlenforschung, Mülheim an der Ruhr, FRG, 30 July–8 August 1984"
"Organized under the auspices of the IUCr Computing Commission"
Includes bibliographies and index.
1. Crystallography – Data processing – Congresses.
I. Sheldrick, G. M. II. Krüger, C. III. Goddard, R.
IV. International Union of Crystallography.
Commission on Crystallographic Computing. V. Title.
VI. Title: Crystallographic computing three.
QD906.7.E4157 1984 548'.028'5 85-15299
ISBN 0-19-855211-4

Printed in Great Britain by
St Edmundsbury Press,
Bury St Edmunds, Suffolk

Preface

The Ninth International School on Crystallographic Computing in Mülheim directly preceded the XIIIth Congress of the International Union of Crystallography in Hamburg, and was attended by 131 participants from 23 countries.

The scientific programme was divided into four sections: data collection and analysis for single crystal and powder samples, the use of data banks, program packages for maxi, mini, and micro computers, and computing methods in protein crystallography. The topics were selected with a view to avoiding duplication of material presented at the IUCr Congress. The programme was organised around the practical work sessions which were scheduled for early morning and late afternoon so as to make the most efficient use of the available computer time. In between, participants attended plenary lectures held in the institute's lecture theatre. The lectures were supplemented by four panel discussions, one on each of the main topics of the summer school, and these were marked by a lively discussion from the auditorium. During the practical sessions the participants worked in small groups (4–5 people) or attended seminars (20–30 people).

Nine computers were available for the practical sessions. They ranged in size from the institute's VAX 11/780 down to a Rainbow personal computer. In addition to the institute's own VAX computers, there were complete systems from Enraf-Nonius (PDP 11/23 plus), Nicolet (Eclipse), and Logis (PDP 11/3). Stoe demonstrated data reduction and analysis of powder diffraction data on a PDP 23 micro. Evans and Sutherland kindly lent a PS 300 colour display unit, and colour terminals were provided by CIS (AED), Datagraph (Rhode & Schwarz) as well as Tektronix. Two personal computers (Professional 300 and Rainbow 100) were donated by DEC. These facilities gave participants an excellent opportunity to become acquainted with the software that was installed at Mülheim for the school. The various program packages were compared using both test data sets and problem structure data brought by the participants. Participants were also able to measure data on the three local four-circle diffractometers under various measuring conditions, including at low temperature. In all, 20 program packages were demonstrated, in several cases for the first time. In addition, scientific posters were presented by O. Adigüzel, J. Hasek, R. Hilgenfeld, G. Lampert, and L. Y. Y. Ma. The lectures or seminars by W. Steigemann (Simultaneous Refinement using X-ray and Energy Information (EREF), and Protein Program Packages), J. Deisenhofer (Molecular Averaging and Solvent Flattening), and H. Göbel (X-ray Powder Diffraction for Applied Crystallography) do not appear in this volume.

The Summer School was organised under the auspices of the IUCr Computing Commission. The programme committee consisted of H. Burzlaff, R. Goddard, K. Huml, C. Krüger, H. Schenk, G. M. Sheldrick, and W. Steigemann; the local organising committee, who carried the brunt of the work, included U. Bartsch, R. Goddard, C. Krüger, and E. Ziegler.

The following organisations provided financial support and/or made computing equipment and peripherals available: CIS GmbH, Viersen; Digital Equipment Corporation GmbH, Munich; Enraf-Nonius GmbH, Solingen and Delft; Evans and Sutherland GmbH, Munich; the International Union of Crystallography (IUCr); Logis Electronic GmbH, Cologne; the Max-Planck-Institut für Kohlenforschung, Mülheim an der Ruhr; Nicolet Instruments GmbH, Offenbach; Rhode und Schwarz GmbH, Cologne; Siber Kikai K.K., Frankfurt; Siemens GmbH, Karlsruhe. A grant from IUCr enabled 15 students (from 13 countries) to attend the Summer School.

The overall success of the Summer School is without doubt in part due to the comprehensive social programme, which included a welcoming reception, a boat trip, and a conference dinner. None of this would have been possible but for the generous financial support from industry and the host institute.

Göttingen
April 1985

G.M.S.

Contents

Program systems for maxi-, mini-, and micro-computers

Data collection and processing for single crystals and powders

REDUCING RANDOM ERRORS IN INTENSITY DATA COLLECTION

E. J. Gabe
Chemistry Division
National Research Council of Canada
Ottawa, Canada K1A 0R6

In order to think about reducing errors, we must first think about errors, and then their causes and their effects. Error is a deviation from 'correctness' or a difference between the measured and 'true' value of a quantity. There is in both these definitions an assumption of the knowledge of 'correctness' or 'truth', which in most cases we do not have.

There is however an apparent difference in principle between the measurement of the length of a steel bar and that of the intensity of a reflection. Most people do not doubt that the steel bar has a 'true' length which is in principle measurable. The intensity on the other hand, merely has a 'most probable' value which can only be expressed in statistical terms. Of course, if we actually try to measure the bar, we find ourselves running into the same types of statistical definitions, but the reasons for this are different. In measuring the bar we are essentially calibrating its length in terms of previously calibrated lengths (rulers, micrometers, wavelengths of light). The need for error estimates arises because of errors in the instruments we use, either random or systematic, and because of the random errors of our own measurements. In the case of the intensity, not only are our instruments imperfect but also the process by which the intensity arises - the production of X-rays or neutrons - is intrinsically random. No matter how hard we try we cannot control the process and even with perfect observers and instruments we can still only measure a 'most probable' value.

In any experiment all these errors, from whatever cause, feed through into the quantities derived from the measurements. Systematic errors in the measurements will produce systematic errors or bias in the results. Random errors in the measurements may or may not produce bias in the results. For example, a random error in the measurement of a crystal face to be used to calculate absorption corrections will bias the intensities; random errors in the intensity measurements themselves however, will probably not introduce bias if the problem is sufficiently overdetermined in a least squares process. For both types of error, the reliabilty of the results, as expressed by their variances, will only be correct if the variances of all the quantities contributing to them are correctly estimated and included. Errors in some measurements are amenable to analytical treatment e.g. counting statistics; others, e.g. face measurement, X-ray tube output fluctuations, crystal movement etc, are less amenable and we resort to devices which we dignify with names like 'instrumental instability factor'. It is however clear that in all cases we should seek to reduce errors in measured quantities - random or systematic - in order to obtain the most accurate results from a given experiment.

Sources of Error in Intensity Measurement

In order to reduce intensity errors we must first find the origins of these errors and then try to eliminate or at least minimize them. By examining the process of measuring intensities step-by-step, we can detect possible problem areas and suggest possible improvements. For this purpose it is convenient to consider the errors as arising from the crystal, the instrument and the experiment. I apologize if many of the topics have little to do with computing, but in the best tradition of 'garbage in - garbage out', I wish to ensure that only the highest quality garbage is allowed in.

1. Crystal

The crystal is the primary source of all intensity information and as such it must be considered a source of error.

It is obvious that good data cannot be obtained from poor crystals and it is important to have a means of assessing the quality of the crystal. The best way to do this is to examine the scan profile in some detail. A precession photograph will detect gross problems, but it is probably insensitive in marginal cases. On the other hand, Laue photos may be too sensitive and cause acceptable crystals to be rejected somewhat indiscriminately. A scan-profile taken with a chart recorder will show gross defects, but will smooth away detail. An oscilloscope however, is an ideal device for this purpose and in my opinion all diffractometers should be provided with such a monitor. Fig. 1 shows details of precession and Laue photos and an oscilloscope trace of a reflection from the same crystal.

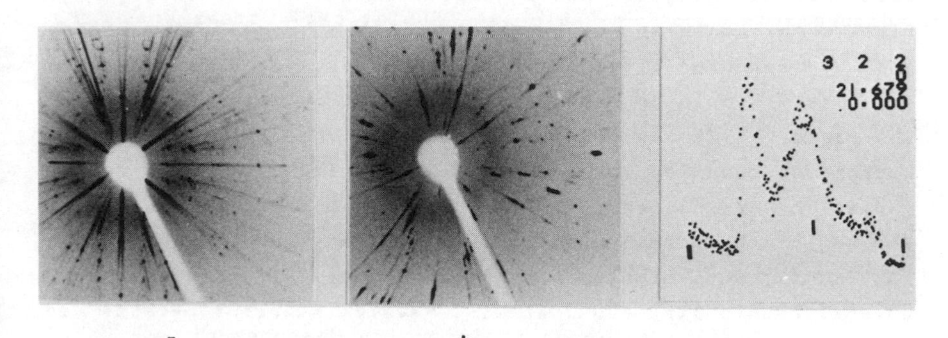

Fig. 1. Precession photo (a), Laue photo (b), and reflection profile (c) of the same poor quality crystal.

If the selected crystal is judged to be unacceptable, there may be others that are better or it may be possible to recrystallize the material. If it is necessary to work with poor crystals it is essential to use scan methods. Increasing the take-off angle may help to smooth the profile, but the effect is mainly cosmetic. Duisenberg (1983) has proposed a far more effective way of dealing with this problem, where the value is calculated for each reflection to give the least dispersed or narrowest profile.

2. Instrument

In this context the instrument is taken as the radiation source, the counting system and the diffractometer. Whereas an unsuitable crystal can be rejected in favour of a better one, most of us do not have that luxury as far as our instruments are concerned. We must therefore try to ensure that they produce the best results possible. This means checking the components with special routines and maybe photographs, using a stable standard crystal. In practice this is usually only done when trouble is suspected, but there should be a routine in the control program to take repeated stationary counts and keep track of short-term and long- and short-term variations. The whole instrument can then be checked in the correct sequence and any problems isolated.

a. Counting System

The counting system consists of the scaler, the pulse-height analyser, the preamplifier and the detector. The first three of these can be checked with a signal generator, and the detector itself with an Fe55 source of 5.9Kev X-rays. The system should be stable within normal counting statistics tolerances. It is equally important to check that the background counting level due to electronic noise and cosmic radiation are acceptably low. With the analyser window set for any of the normal wavelengths used, one should expect 5 to 10 counts per minute background rates. If this climbs to several counts per second it will have deleterious effects on the data, particularly the weak reflections, and will show up as increased numbers of insignificant reflections and high standard deviations. Curing problems of this type will normally need expert electronic service, and the causes can be anything from noisy components to misplaced or unshielded wires.

Once the counting system is functioning properly one can proceed to the next step in the sequence.

b. Radiation Source

It is obvious that the intensity of the incident beam must be acceptably constant. This is easily checked using the same routine as for the counting system. In this case a stable standard crystal (ruby) should be used as a sampling device. Unacceptable results will normally require maintenance, the form of which depends on the particular problem. In the case of neutrons and X-rays from synchrotrons a cry for help is appropriate. For X-ray tubes the corrective action is normally the responsibility of the operator. The problem can be isolated to the generator or cable or tube by swapping components and then taking appropriate action.

If you are in the habit of looking for trouble, it can be

found very easily by taking pinhole photographs of the source, either directly or through the monochromator. The intensity distributions found are often very non-uniform, but beyond swapping components there is little one can do to improve the situation. The uniformity of the beam at the crystal can be checked by traversing a fine pinhole (25 micron Pt used in electron microscopy) across the beam with the translation motion of a goniometer head. The distributions will be quite different in the vertical and horizontal directions if a monochromator is being used, but should be quite smooth (Coppens et al., 1974). Limits to the useable crystal size can be established from these scans, which normally show a minimum plateau dimension of about 0.5 mm.

This is a good point at which to check the pulse-height analyser settings. The upper and lower voltage levels are particularly important. It is very easy to forget to reset the analyser after a tube change, Mo to Cu for example, and this will degrade the data very badly.

c. Diffractometer

Again, it is obvious that a diffractometer must be reasonably well aligned in order to yield good data. Many users veiw the alignment process as being complex, time-consuming and vaguely unsatisfactory. This need not be so and there is no reason why fear of alignment should prevent a tube change to optimize data collection conditions. Alignment for intensity measuring purposes, can be done 'optically', with a few pinhole apertures, in half an hour. This can then be checked with a standard crystal using a multiple setting routine (Hamilton, 1974) and minor adjustments may be made. Generally these will not be necessary, but the zero values of the angles are found with this procedure and it is convenient if the control routine uses these values. Alignment for the measurement of precise lattice parameters is more critical, but is only necessary for the most accurate work, roughly better than 1 part in 20000. This precision is unattainable with the average organic crystal and one should not strive after such accurate alignment if it is not needed.

With a satisfactorily aligned instrument one should check

that the mechanics of scanning produces consistent intensities. A continuous scan must be driven by a synchronous motor in order to give reproducible results, and a step scan must have a setting error small enough to prevent the introduction of statistical noise.

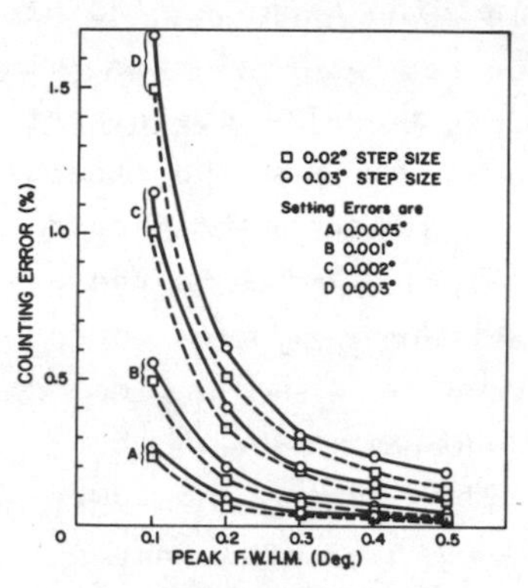

Fig. 2. Integrated intensity errors caused by random setting errors during step scanning. The size of the setting error is indicated as A, B, C or D.

Fig. 2 shows that intensity errors caused by step scanning increase with setting error and also as the number of steps within the peak decreases. This is to be expected, but the size of the errors could be unacceptable for the most accurate work. It is regrettable that no currently available commercial diffractometer provides a continuous scan facility. The simplest solution is to subdivide each step into a series of substeps in order to sample the peak envelope more frequently and this should be implemented either in hardware or through software.

3. Experiment

Under this heading are included the operations which must be performed in order to measure intensities and the choice of conditions to minimize the errors. There is no description of Laue group and space group determination because these are dealt with elsewhere and also because any errors made can hardly be considered 'random'.

a. Crystal Mounting

The crystal mount should aim to maintain the crystal in the beam in a stable manner, with the minimum amount of extraneous material in order to reduce unnecessary background scatter. Jamming the crystal in a capillary tube gives good stability but risks damaging the crystal and increases the scatter from the extra glass. If the crystal is mounted on a fibre a minimum amount of glue should be used and the glue should be thoroughly hardened before starting any work. A fast curing epoxy glue has good properties for this purpose. It is extremely sticky so that a minute bead on the very tip of the fibre can be used, it does not form a dry skin as most solvent glues do and it allows a reasonable length of time for the crystal to be manipulated under the microscope. Even though it is a fast-curing glue, typically 5 mins, it is nevertheless a good idea to allow several hours for thorough setting to eliminate unwanted movement.

Such movement can also occur between the fibre and the metal mounting pin or in the goniometer head itself, which should be as simple as possible i.e. it should not have any angular adjustments. Do not mount the fibre so that it is bent or stressed and try not to move the goniometer sledges once the lock screws have been tightened. In either of these cases the stored stress will be relieved somehow - with accompanying rotation or translation - during the next few days. Such movements occur with annoying frequency and it is essential to be able to detect and correct them before they cause significant errors. Again an oscilloscope is an excellent device for visually monitoring reflection profiles. A correction routine should be set up to come into operation when the profile has drifted beyond preset limits. Translational movements require human intervention to recentre the crystal, but fortunately most movements are rotational and can be corrected automatically.

b. <u>Crystal Face Measurement</u>

In order to calculate absorption corrections by Gaussian quadrature or by the analytical method, it is necessary to describe the shape and size of the crystal. The measurements involved can introduce significant errors into the final corrected intensities.

Each face may be described in terms of two orientation angles and a distance from the centre of the crystal. These measurements are most conveniently performed on the diffractometer, but several methods have been proposed for both 'in-situ' and remote measurement with microscopes and optical goniometers. Errors will occur whichever method is used. If the faces can be assigned h, k, l values, the angular errors will be negligible, though usually this is not the case and both angular and distance errors occur. To minimize these errors the crystal should be measured several timnes - preferably by different people - and any differences resolved. Attempts should be made to account for rounded corners and edges by including small 'faces'. These measurements are very subjective and it is essential to check that the shape represented really does look like the shape of the crystal. This is best done at a graphics terminal while the crystal is still available for remeasurement and the routine should allow corrections to be made. Relatively small changes can affect the shape of the crystal dramatically.

The mismatch between the 'real' and measured shapes of the crystal can cause integration points or corners of tetrahedra to be placed outside the real crystal (see Fig. 3), and as the value increases large random errors can occur. Howard Flack may have more to say about this. Calculation errors on the other hand can be made negligibly small, and thus in order to improve absorption corrections one should strive for the best possible representation of the crystal.

c. Choice of Conditions

There are several options which might be considered to enhance the quality of intensity data. They are increased incident beam intensity, increased counting times, choice of wavelength, proper collimation and the use of low temperatures.

The use of high intensity or rotating anode tubes does not of itself guarantee significantly better data. An increase of beam intensity by a factor N should produce a $\sqrt{N}$ improvement in the signal-to-noise ratio, which is not to be despised. Other factors such as cost, convenience and increased instability may outweigh this advantage however. If the crystals under investigation have a limited life, then it may be desirable to trade beam intensity against data collection time, as for protein work. Most laboratories use standard sealed X-ray tubes and it is our experience that they perform best when they are run at about 75% of their maximum rated output. Under these conditions the output is extremely stable over long periods of time and the tube life is very long (>20000 hours).

The same improvements in signal-to-noise ratio can be achieved by increasing counting times, but the total time involved can become extremely long - a factor of N for even a modest improvement of $\sqrt{N}$. There are advantages in using the extra time to collect equivalent unique sets, because with a randomly oriented crystal multiple reflection effects will be different for each equivalent. Also the time to collect one complete set is kept to a minimum thereby reducing any scaling corrections and instability efects..

The greatest improvements in intensity data can be made by improving the quality of the weak reflections. This can be achieved by improving the signal-to-noise ratio by proper collimation, use of appropriate wavelength, monochromatization and the use of low temperatures.

Collimator apertures can be tuned to the particular crystal in order to reduce background intensity. Care should be taken not to be too drastic and clip the actual peaks. Crystal movement should be carefully monitored during the data collection if the collimator apertures are small.

In general the longer wavelengths (Cu, Co) are intrinsically 'cleaner' than the shorter ones (Mo, Ag), but if one is forced to use short wavelengths because of absorption the radiation can be cleaned up with a monochromator. Scattering power increases with wavelength and longer wavelengths should be used for light atom crystals with low absorption. The crystal size can also be increased in this case.

The foregoing techniques all seek to reduce the background, but one can also improve the signal-to-noise ratio by increasing the signal. Lowering the temperature will do this, but will not necessarily improve the signal-to-noise ratio in an effective way. The experiment is inherently less stable under these conditions and this instability may cancel out any gains made by reducing thermal motion. These effects are particularly noticeable in heavy atom materials, where the increased scattering from the heavy atoms tends to dominate the low temperature data, particularly at high angles, and the accuracy of the light atom contributions may not be significantly improved. For light atom structures however, the use of low temperatures usually improves the data significantly.

d. Orientation Matrix

The orientation matrix used for data collection is often quite crudely determined from a small number of reflections, perhaps only three. This is often sufficient as long as scan techniques are used, but can give rise to some problems. The three reflections define a segment of reciprocal space and reflections within that segment are usually – though not necessarily – well centred. Reflections in other segments are often less well centred and frequently require further alignment of the crystal before data collection can proceed. To cure this, one should either use more reflections or the original small number plus their Freidel equivalents. The final cell parameters should be derived from a matrix obtained with high-angle reflections selected after the intensity data has been measured (Clegg, 1984). All symmetry equivalents may be used, which will distribute the reflections in reciprocal space and the $\alpha1/\alpha2$ splitting is usually dealt with by measuring $\overline{\alpha}$ or $\alpha1$ alone. In order to obtain reliable cell parameters it is necessary to use between 40 and 80 reflections and the most accurate values are derived from 2θ data alone.

e. Multiple Reflection

Multiple reflection is present in all intensity data and cannot be completely avoided. Renninger (1937) recognized it as a diminution of some intensities, usually strong ones (aufhellung), and an enhancement of others (umweganregung). The effect becomes more noticeable when the average temperature factor is small, because necessarily the temperature factor associated with multiple reflection is at least twice that for single reflection. The energy subtracted from strong reflections reappears in other, usually weaker reflections. The effect of taking intensity from the strong reflection is minor, but the redistribution to the weaker reflections can have disastrous effects on their intensities. Fortunately the phenomenon is very dependent on orientation and experimental attempts to cope with the problem have centred on reorientation of the crystal. The simplest procedure is to measure several equivalent reflections and only accept consistent measurements at data reduction time. Another method is to measure the same reflection at slightly different ψ values and again look for consistency. This has the advantage that the process can be repeated until consistency is obtained though both of these methods can be very time consuming. Another method proposed by Coppens (1968), is to calculate the ψ value for a reflection, at which the double reflection effects for a list of strong reflections will be a minimum.

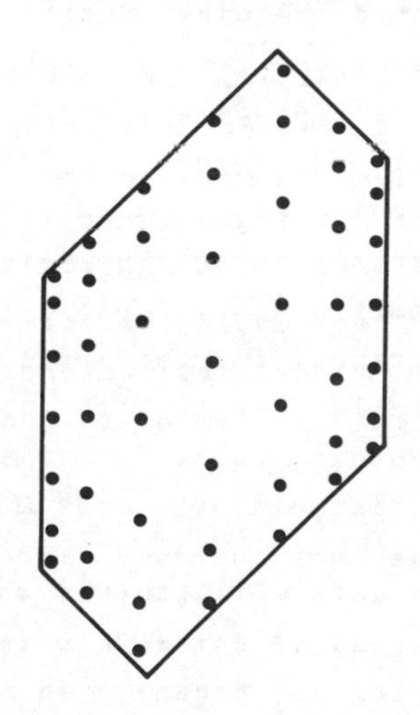

Fig. 3
Distribution of integration points in a 2-D shape. Rounding of the corners will remove some points.

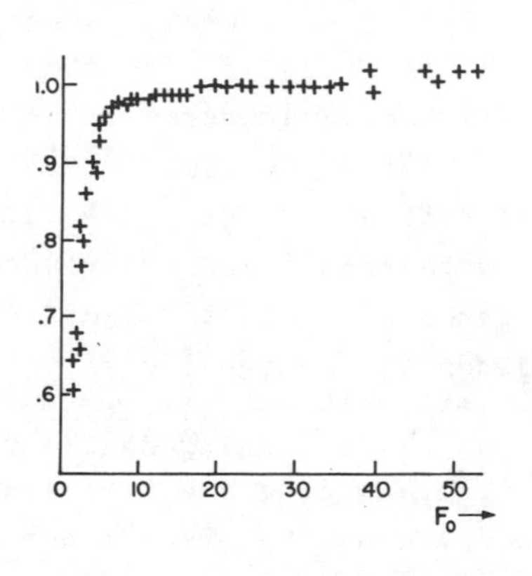

Fig. 4
Plot of <Fc/Fo> for epidote. Clearly, when Fo is small the measurement is too large.

None of these methods is ideal, but for inorganic crystals multiple reflection effects are significant (see fig. 4). The first method is the simplest, but care should be taken to avoid rejecting valid measurements. It is almost impossible to detect the diminution of the strong to medium reflections by statistical tests, but the enhancement effects on weaker reflections can be detected, as seen from the following table.

	h	k	l	I(hkl)	I-(hkl)	h	k	l	I(hkl)	I-(hkl)	Fo	Fc
	1	3	19	229	257	3	-4	19	159	187	4.53	4.34
	2	1	22	34	17	1	-3	22	35	70	2.03	1.24
	5	0	2	538	519	0	-5	2	317	355	5.39	4.57
	5	3	2	37	41	3	-8	2	15	-35	1.59	0.71
	7	1	12	9	14	1	-8	12	29	26	1.66	0.98
	8	0	2	49	87	0	-8	2	20	12	1.94	0.90
	8	0	8	127	110	0	-8	8	156	175	3.99	3.74
*	2	0	5	639	674	0	-2	5	-3	23	-	0
*	2	0	17	9	-8	0	-2	17	120	97	-	0
*	3	0	9	12	25	0	-3	9	118	134	-	0
*	7	0	7	20	15	0	-7	7	250	178	-	0

Multiple Reflection Intensity Data from a Ruby Crystal
Equivalent reflections h, k, l; -h, -k, -l; k, -h-k, l; -k, h+k, -l.
Systematic absences marked *. Mean B value 0.3A^2, R(F)=0.011.

Several points can be made. All the intensities are weak (including systematic absences); in every case Fo is greater than Fc, implying that the lower intensities are more correct; the matching intensities occur in pairs of h, k, l and -h, -k, -l, meaning that Friedel reflections suffer the same multiple reflection effects. A discriminating test would therefore have to detect weak reflections, group Friedel reflections together, detect significant statistical differences and favour the lower values.

f. <u>Reflection Measurement Order</u>

Minimizing the time taken to measure data will contribute to reducing the overall errors and one method of doing this is to measure reflections in the most efficient order. This means selecting the index corresponding to the longest axis to be the fastest varying, and also to measure up and down the rows of this index in a zig-zag manner. This minimizes the time spent driving between reflections, and will shorten the total time appreciably for very anisotropic cells. If Friedel reflections are to be collected it is best to measure them as close together as possible, but obviously not one after the other. A reasonable compromise is to measure one or two rows of h,k,l reflections and then the -h,-k,-l equivalents. This puts the reflections reasonably close in time while minimizing the extra driving time.

g. Profile Analysis and Counting Statistics

The safest simple method to collect intensity data is with a background-peak-background (B-P-B) scan. In this method if the backgrounds b1 and b2 are observed at each end of the scan for 1/mth of the time taken to measure the peak P, then

$$Inet = P - m(b1 + b2) \quad \text{and} \quad \sigma^2(Inet) = P + m^2(b1 + b2)$$

based on so-called counting statistics. Analogous expressions can be derived, no matter how the intensity data is collected, whether it be B-P-B, some form of statisitical profile treatment, profile fitting techniques or restricted scan methods. An excellent survey of these methods has been given by Clegg (1984) and I do not intend to go over the details again. It is important to realize however, that there are many methods to improve on the basic B-P-B technique. All such methods measure the peak as a series of steps, sometimes limited in number, in a profile scan and attempt to squeeze as much information as possible from these profile points. By improving the signal-to-noise ratio Inet/σ(Inet) they allow one either, to achieve a given accuracy in a shorter time, or a greater accuracy in a given time.

Profile treatment techniques (Lehmann & Larsen, 1974; Tickle, 1975; Grant & Gabe, 1978; Rigoult, 1979) attempt to find the point on either side of the peak where the background begins, reducing the amount of background noise under the peak and thus improving the statistics. Also, as more of the sides of the profile are included in the background the scans can be narrower and the background measurements shorter. Profile fitting methods (Diamond, 1969; Clegg, 1981; Oatley & French, 1982) attempt to fit a learned profile to the profile points for each reflection. This incorporates knowledge of previous reliable reflections into each measurement and thus removes statistical noise. The effects are similar to profile treatment but are perhaps more reliable. With any of these methods the improvement is particularly noticeable on weak reflections, which is exactly where it is needed most. The main aim of restricted scan methods is to increase the speed of data collection, while preserving as much accuracy as possible. A single, hopefully peak-top, measurement is the simplest and most dangerous of these methods. Other techniques (Wyckoff et al., 1970; Bassi, 1976; Hanson et al., 1979) use a small number of profile points and either perform a peak integration or obtain the intensity using a previously learned function. Background values are often obtained from a table established at the start of, or during, the experiment.

These methods are often applied after the fact i.e. as off-line techniques, with the data being stored on disc or tape. In my opinion the processing should be done on-line, even at the cost of some small delay for computing. In this way any problems can be immediately resolved by repeating the measurement. I also feel, as I have already indicated, that it is extremely important that the peak profiles be continuously monitored on some form of display unit, so that the operator can check peak shape, peak width, peak positioning and profile limits, quickly and easily during the experiment.

Conclusion

I have tried to outline, in general terms, the steps necessary to try to reduce random errors in intensity measurement and to ensure that a diffractometer will yield good data. It is not, however, an automatic prescription, as there is no single path to high quality intensity data. The techniques should be used with care and discrimination, but regrettably this is often not the case, and instruments are set up and left to run for months or years with little checking or change of technique.

Assuming a well aligned and properly functioning instrument, the use of profile analysis offers the most significant improvements for intensity data collection. When we installed our profile analysis routine, the data collection rate effectively doubled, with no apparent loss of precision, and I agree wholeheartedly with Clegg (1984) when he says "It is a tragic fact that the majority of diffractometer time in the world today is being inefficiently used". Allan Larson once said to me "The trouble with writing software is that people use it!", and this is a real prblem. People use the software they have and the fact that no commercial package offers a profile analysis routine means that most people do not use one. I hope that the situation will soon be remedied even if fewer diffractometers are sold because of increased efficiency.

References

Bassi, G. (1976). Crystallographic Computing Techniques, Ed. F.R. Ahmed, Munksgaard, Copenhagen, p. 197.

Clegg, W. (1981). Acta Cryst. A37, 22-28.

Clegg, W. (1984). Proceedings of the Kyoto Summer School.

Coppens, P. (1968). Acta Cryst. A24, 253-257.

Coppens, P., Ross, F.K., Blesing, R.H., Cooper, W.F., Larsen, F.K., Leipoldt, J.G. & Rees, B. (1974). J. Appl. Cryst., 7, 315-319.

Diamond, R. (1969). Acta Cryst. A25, 43-55.

Duisenberg, A.J.M. (1983). Acta Cryst. A39, 211-216.

Grant, D.F. & Gabe, E.J. (1978) J. Appl. Cryst. 11, 114-120.

Hamilton, W.C. (1974). Int. Tables for X-ray Crystallography, Vol. IV, p. 282.

Hanson, J.C., Watenpaugh, K.D., Sieker, L. & Jensen, L.H. (1979). Acta Cryst. A35, 616-621.

Lehmann, M.S. & Larsen, F.K. (1974). Acta Cryst. A30, 580-584.

Oatley, S. & French, S. (1982). Acta Cryst. A38, 537-549.

Renninger, M. (1937). Z. Phys. 106, 141-176.

Rigoult, J. (1979). J. Appl. Cryst. 12, 116-118.

Tickle, I.J. (1975). Acta Cryst. B31, 329-331.

Wyckoff, H.W., Tsernoglou, D., Hanson, A.W., Knox, J.R., Lee, B. & Richards, F.M. (1970). J. Biol. Chem. 245, 305-328.

AVOIDANCE, DETECTION AND CORRECTION OF SYSTEMATIC ERRORS IN INTENSITY DATA

H.D.Flack

Laboratoire de Cristallographie aux Rayons X, Université de Genève,
24, quai Ernest Ansermet, CH-1211 Genève, Switzerland

Introduction

The subject matter which could be covered by the title of this presentation is immense. I attempt to highlight some of the aspects which have not been previously covered in sufficient detail in the computing summer schools or require restatement. The section on Serial Correlation and Correlated Residuals concerns a problem which is partially unsolved. It is hence offered as food for thought.

Axial Polar Dispersion Error

Ueki, Zalkin and Templeton (1966) in their study of thorium nitrate pentahydrate found that an error of about 0.05A is made in the position of the thorium atom as determined by least-squares refinement if the anomalous dispersion contribution of Th (f" = 9) is neglected in the calculation of the structure factors. This type of error only occurs under the following conditions:

a1) anomalous dispersion has been neglected
b1) the non-centrosymmetric space group is axially polar i.e. has a free origin in 1, 2 or 3 directions
c1) only one asymmetric region of the Laue group of reflections has been measured.

The shift of the Th atom is along the axially polar direction and in such a way as to compensate for the neglected anomalous dispersion. It may seem contradictory that the Th atom was used in this study to define the origin but in practice the least-squares refinement just moves the other atoms in block by the same amount and in the opposite direction. The shift of the atoms compensates so well for the neglected f"s that there is little to choose between the R factors for refinement with and without anomalous dispersion. In the literature this type of error has come to be known as a polar dispersion error but because of the confusion in the crystallographic use of the word polar (Waser,1974; International Tables for Crystallography Vol.A, 1983) I prefer to add the word axial for clearness. Cruickshank and McDonald (1967) analysed this situation in more detail and give in Fig. 1 typical coordinate errors arising when conditions a1, b1 and c1 are fulfilled.

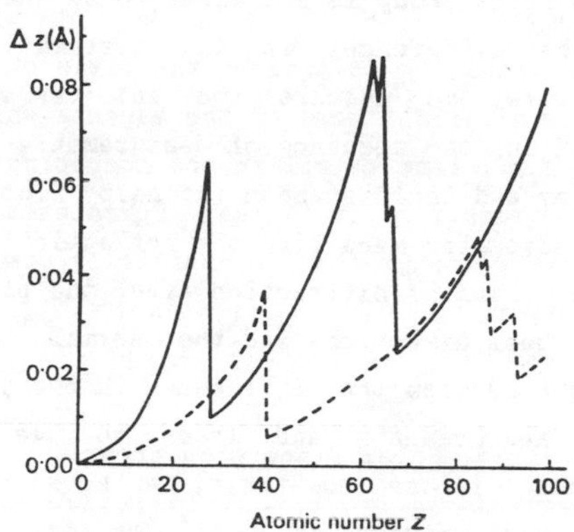

Fig. 1. Typical coordinate errors due to neglect of $\Delta f''$ in polar space groups. Full line: Cu $K\alpha$ radiation. Broken line: Mo $K\alpha$ radiation.

With the generalised use of anomalous dispersion contributions in structure-factor least-squares calculations, it could have been thought that axial polar dispersion errors were dead and buried. In the words of Wilson (1975) " all that really needs to be said is - Do the right calculation and get the right answer". But no. Templeton, Templeton, Zalkin and Ruben (1982) found in comparing their own results on Pr, Sm and Gd edta salts with values for Dy edta obtained by Nassimbeni et al (1979), that the Dy position was wrong by about 0.07A. In this case it would seem that although anomalous dispersion had not been neglected, conditions b1 and c1 were fulfilled and that structure refinement had been carried out in the wrong polarity i.e. in an orientation related by a centre of symmetry to that of the real crystal. It is easy to show that this will result in a coordinate error twice that of Fig. 1. To conclude, if

a2) anomalous dispersion had been included

b2) the non-centrosymmetric space group is axially polar
c2) only one asymmetric region of the Laue group has been measured
d2) the two polarities have not been tested

then it may be that an axial polar disperion error of twice that of Fig. 1 has been made.

What can be done to eliminate this type of error? In order to overcome condition c2, a cautious approach would be to measure both reflection and anti-reflection if the space group is not known to be centrosymmetric. In order to minimize the differences in the systematic effects on the intensities, one is advised to measure the anti-reflection immediately after the reflection in the sequence of measurements (thus reducing the effects of crystal decay and incident beam intensity fluctuation). Clegg (undated) has further advocated measuring the reflection and antireflection pair using the same plane of diffraction (i.e. the plane containing the incident and reflected beam directions and the normal to the reflecting plane). This technique reduces the differences in the path lengths in the crystal for the two measurements and leads to identical absorption, extinction and thermal diffuse scattering effects for a crystal of centrosymmetric shape. With regard to condition d2, easily applicable techniques are now available for testing absolute structure. Dr.P.G. Jones describes these techniques in his seminar with special reference to the problems of crystal chirality. I must however emphasize the importance of testing the absolute structure in any non-centrosymmetric space group whether one is interested or not in the crystal chirality or polarity in order to obtain unbiased atomic postional and thermal parameters.

Absorption

Refinements on intensity data from an absorbing sample for which no correction has been made will suffer from temperature factors being too small or even negative and from poor agreement factors. The atomic positional parameters will be very little altered due to the lack of an adequate absorption correction. The above two statements contain the justification for both carrying out and not carrying out an absorption correction! The choice of programs available for absorption correction is considerable and some thought should be devoted to choosing one of those most suited to the study in question. A detailed overview with complete references of the available techniques can be found in the proceedings of the Kyoto Summer School (Flack,1983).

Absorption Correction by Integration

The classical technique of absorption correction involves measuring the external shape of the crystal and evaluating the transmission coefficient by numerical quadrature (integration). The most difficult part of the procedure is in the measurement of the shape of the crystal. Those studies or programs which allow the dimensions of the crystal to be optimized based on intensity measurements of equivalent reflections or azimuthal scans have often found that very small or even insignificant changes in crystal dimensions considerably improve the internal consistency of the correction. The most commonly used method of numerical quadrature for absorption correction is that based on a Gaussian grid. This grid has a maximum density of grid points near the surface of the crystal. For a homogeneous sample the value of $\exp(-\mu t)$ varies rapidly near the surface of the crystal and more slowly in the interior of the crystal. One can thus see that for a homogeneous sample a Gaussian grid will be superior to an isometric grid. In any situation where the sample is not homogeneous within the integration volume, the Gauss method becomes less interesting and one should consider the use of a Monte Carlo method where the grid points are distributed at random. Also of particular interest are the modern adaptive techniques which try to accumulate the grid points in those regions where $\exp(-\mu t)$ is varying most rapidly based on an analysis of the function itself.

It is also possible to express the transmission factor of a convex polyhedral crystal in an analytical form by cutting the integration volume up into suitable polyhedra and tetrahedra. Although analytically exact the method suffers from numerical problems of rounding. The calculation time is however independent of the value of μ unlike the numerical quadrature methods where the number of grid points and hence the calculation time must increase with μ.

Semi-Empirical Absorption Corrections

These methods attempt to make an absorption correction by using information in the intensity data itself rather than by measuring the geometry of the diffraction experiment. The intensity data will be in the form of measurements of a selected set of symmetry equivalent reflections and possibly of azimuthal scans of these. Some function is chosen with free parameters and these free parameters are adjusted to improve the equivalence of the symmetry equivalent measurements. Once the transmission factor has been obtained in this form, it can be used to apply a correction to the whole data set. In the most general case the function would have four angles as independent variables and it is common to simplify by taking some physical approximations. However the choice of suitable functions and hence significant independent variables may be left to the choice of an adaptive fitting scheme related to the adaptive numerical quadrature mentioned above. In this type of scheme candidate functions in the representation of the transmission coefficient are tested sequentially. If the new function makes a significant reduction in the sum of squares of residuals of corrected measurements it is accepted for inclusion in the transmission factor representation. Otherwise it is temporarily rejected and retested later. By the trick of making linear transformations of the candidate functions the equivalent of a diagonal least-squares normal equations matrix is maintained which may be updated sequentially with the accepted transformed functions.

One important limitation of fitting symmetry equivalent intensity data is that the pure dependence on theta of the transmission factor is unobtainable. This is because symmetry equivalent measurements all occur at the same value of theta.

There is one semi-empirical method available which uses a functional representation of the transmission factor but obtains the values of the coefficients by minimizing the differences between F_{obs} and F_{calc} from a model. This method must suffer a larger atomic parameter bias than those not involving F_{calc}.

Incident Beam Inhomogeneity

X-Ray beams obtained by diffraction from a monochromator are spatially inhomogeneous. The form of the beam arriving on the sample may be seen by taking photographs or by carrying out intensity measurements through a pin-hole. The results of Harkema et al (1980) are most interesting. These workers have incorporated an approximation to this intensity distribution into their Gaussian quadrature absorption program. They assume that the intensity distribution may be described as the product of two parts, one of which is dependent only on x and the other only on y. It is a good approximation to take the x dependence as a constant and the y dependence as a Gaussian. The absorption correction program is set up so that the planes of the Gaussian grid points are taken to be perpendicular to the y axis. The grid points may thus be weighted according to the intensity incident upon them. Such a scheme will work perfectly well even for a non-absorbing crystal. An interesting corollary is that for azimuthal scans the effect of absorption and inhomogeneity are out of phase by 90 degrees.

Almost Spherical Crystals

For those working in the field of high accuracy measurements be it for electron density, thermal vibration or whatever, it is often advantageous to produce spherical or near spherical crystals by grinding very gently in a mill. In this way one reduces the problems of absorption and incident beam inhomogeneity. Moreover the sphere is a shape for which the currently used theories of secondary extinction were established. The grinding process may produce a damaged surface layer which is more ideally imperfect (i.e. diffracts kinematically) than the interior of the sphere which would suffer a larger secondary extinction effect (Boehm,Prager & Barnea,1974). This is an example of an inhomogeneous mosaic structure. It is desirable in this case to render the mosaic homogeneous by chemical etching to remove the damaged surface layer. Should this be unsuccessful, the inhomogeneity should be treated in any secondary extinction correction applied to the data. Le Page & Gabe(1978) assume that the crystal volume is made up of a non-extinct fraction x diffracting with the kinematical intensity and an extinct fraction 1-x which suffers secondary extinction according to one of the usual theories. The total diffracted intensity is taken as the sum of the intensities of the extinct and non-extinct part. The volume fraction x is treated as a parameter in the least-squares refinement of the data. One of the excercises is intended to make you discover a more complex example.

Crystals produced by grinding are frequently not exactly spherical in shape. It is therefore interesting to assess the magnitude of the intensity variations between equivalent reflections or in azimuthal scans caused by departure from sphericity. Working the other way round, this type of information may be used to determine the maximum permissible departure from sphericity, as measured by the relative root mean square variation in the radius, for a chosen maximum variation in the equivalent intensities. Fig.2 shows typical curves giving the maxmimum permissible radius variation as a function of μR for various values of intensity variation. These and other curves are to be found in Vincent & Flack (1979) and Flack & Vincent (1979).

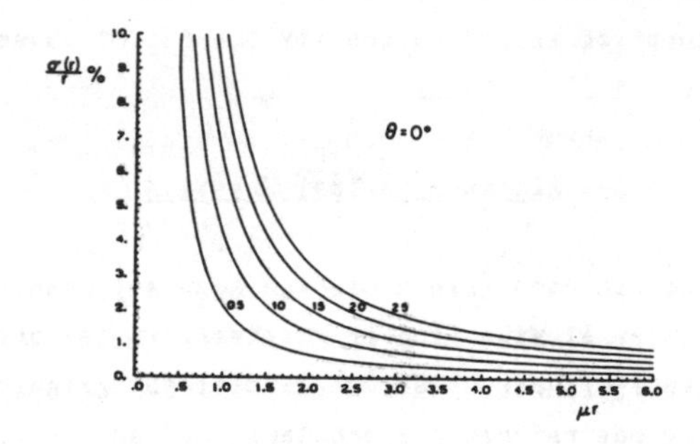

Fig.2 Curves of % radius variation against μR at five values of % intensity variation.

Serial Correlation and Correlated Residuals

Systematic effects remaining after a least-squares refinement may be shown up by the use of the Durbin-Watson d statistic which is defined as follows:

$$d = \text{Sum} (D_i - D_{i-1})^2 / \text{Sum } D_i^2 \;:\quad D = F_{obs} - F_{calc} .$$

This d statistic is very widely used by econometricians in the study of time series but seems to have been completely neglected by crystallographers. Notice that the value of d depends upon the order into which the data are sorted and may thus be used to analyse different possible systematic effects. For example if the data is presented in the chronological order of measurement one would expect to detect problems with the variation of the standard reflections (Flack,Vincent & Vincent,1980), or if in ascending order of sin(theta)/lambda one is seeing temperature factor, extinction correction or atomic scattering factor problems. As both the observed and calculated structure factors enter into the evaluation of d, it is clear that the perturbations may be due either to the model or to the intensity data.

One can see from the definition of d, that if no systematic effects remain d will take a value of about 2. If consecutive values of delta F tend to have the same sign then d will take values between 0 and 2 and alternating signs of the delta F's lead to d between 2 and 4. Limited tables of significance values (Durbin & Watson, 1950,1951,1971), approximate normal significance values (Theil & Nagar, 1961) and a computer program for the calculation of the moments of the probability distribution function of d in general cases (L'Esperance, Chall & Taylor,1976) are available.

It is possible to show that where there is correlation among the residuals, i.e. where one obtains a value of d that is significantly different from 2, then although the parameter estimates will be unbiased their estimated standard deviations can be very greatly in error and may either be under or overestimated. One then has the choice between identifying the source of systematic error (and correcting for it) or having meaningless estimated standard deviations.

The time series such as those of Durbin and Watson are treated by a model of first-order Markov auto-regressive scheme where the statistical disturbance of the Nth. measurement is taken as a linear function of that of the N-1 th. measurement. In the realm of physical measurements, gas electron diffraction has some similarities to this case, as points on a spectrum very close together will suffer correlated disturbances due, for example, to incident beam fluctations (Morino,Kuchitsu & Murata,1965; Murata & Morino,1966). In the least-squares refinement of such data one takes account of the non-diagonal nature of the variance-covariance matrix of the intensity observations by using a non-diagonal (but banded) weights matrix (this being of course the inverse of the variance-covariance matrix of the residuals). These two cases are examples of systematic effects on the statistical fluctuations.

The Rietveld profile refinement for neutron or X-Ray diffraction from crystalline powder samples has been the scene of claim and counter-claim as to whether the estimated standard deviations of the atomic parameters are underestimated or not. It would certainly be of the greatest help if the value of d were to be calculated in all Rietveld refinements and quoted in publications. As the interval between steps is fairly large, the statistical fluctations of adjacent observations are not correlated. So any correlation of the residuals must come from deficiences in the model. Prime candidates in this case would be the profile function and the structure factor which combine in a highly non-linear manner. Rollet(1983) has used a non-diagonal weights matrix for Rietveld refinement.

I am aware of two situations where correlated residuals may be producing incorrect estimated standard deviations in single crystal X-Ray diffraction experiments. The first is in the refinement of anisotropic extinction parameters where there are plentiful azimuthal scan data. The decomposition of the structure factor into a geometrical and an extinction part is very similar to the Rietveld situation. For example our very high quality measurements on TiC gave 0.25% for a conventional R factor, 1.73 for the goodness of fit and 0.99 for the d statistic (data ordered on theta). All this for over 3000 observations on 122 unique reflections. The other situation is the determination of absolute structure. In this case some preliminary simulations on generated data (which had no systematic errors) gave estimated standard deviations on an absolute structure parameter too large by a factor of 1.15 for a value of a pseudo d of 2.32 . However for the few real data sets where we have calculated the pseudo d, small values lower than 0.5 have been obtained.

References

Boehm,J.M.,Prager,P.R. & Barnea,Z. (1974). Acta Cryst. A30, 335-337.

Clegg,W.(undated). Cited by Jones,P.G.(1984). Acta Cryst. A40, In the press.

Cruickshank,D.W.J. & McDonald,W.S.(1967). Acta Cryst. 23, 9-11.

Durbin,J & Watson,G.S.(1950). Biometrika 37, 409-428.

Durbin,J & Watson,G.S.(1951). Biometrika 38, 159-178.

Durbin,J.& Watson,G.S.(1971). Biometrika 58, 1-19.

Flack,H.D.(1983). Proceedings of the Kyoto Summer School. O.U.P.

Flack,H.D. & Vincent,M.G. (1979). Acta Cryst. A35, 795-802.

Flack,H.D.,Vincent,M.G. & Vincent,J.A.(1980). Acta Cryst. A36,495-496.

Harkema,S.,Dam,J.,Van Hummel,G.J. & Reuvers,A.J. (1980). Acta Cryst. A36, 433-435.

International Tables for Crystallography (1983). Vol. A. pp 782-783. Dordrecht: Reidel.

L'Esperance,W., Chall,D. & Taylor,D.(1976). Econometrica, 44,1325-1326.

Le Page,Y. & Gabe,E.J.(1978). J.Appl. Cryst. 11, 254-256.

Morino,Y.,Kuchitsu,K. & Murata,Y.(1965). Acta Cryst.18,549-557.

Murata,Y. & Morino,Y.(1966). Acta Cryst. 20, 605-609.

Nassimbeni,L.R., Wright,M.R.W., Van Niekerk,J.C. & McCallum,P.A.(1979). Acta Cryst. B35, 1341-1345.

Rollet,J.S.(1983). Proceedings of Kyoto Computing School, O.U.P.

Templeton,L.K., Templeton.D.H., Zalkin,A. & Ruben,H.W.(1982). Acta Cryst. B38, 2155-2159.

Theil,H. & Nagar,A.L.(1961). J.Am.Stat.Assoc.56,793-806.

Ueki,T., Zalkin,A. & Templeton,D.H.(1966). Acta Cryst. 20, 836-841.

Vincent,M.G. & Flack,H.D. (1979). Acta Cryst. A35,78-82.

Waser,J.(1974). Acta Cryst. A30, 261-264.

Wilson,A.J.C.(1975). In Anomalous Scattering, edited by S. Ramaseshan & S.C. Abrahams, pp 325-331. Copenhagen: Munksgaard.

DATA COLLECTION: AVOIDING BLUNDERS

William Clegg
Institut für Anorganische Chemie der Universität, Tammannstrasse 4,
D-3400 Göttingen, Federal Republic of Germany

Even if random errors are minimised and systematic errors avoided or corrected, a diffractometer data set can be ruined as a result of blunders committed by either the user or the control software. Some possible mistakes are discussed here, together with suggestions for avoiding them. The seminar is based largely on examples. Although the use of suitable software can reduce the likelihood of falling into a trap, and the user-program interface (input and output) plays an important part too, ultimate responsibility must lie with the user.

1. Errors in unit cell determination

Thorough prior photographic investigation of cell and space group will avoid many errors, but is not always possible. Blind reliance on automatic routines for reflection search and cell/orientation determination is particularly dangerous. For the background to cell determination and a summary of diffractometer methods, see Clegg (1984c) and other previous Summer School lecture notes.

(a) If the reflections used for cell determination all belong to a special subset, an incorrect cell is likely to be found. For example, if all reflections have even k, the b-axis will be halved in length; if all reflections have even h+k, a C-centred cell will be indicated. The danger is particularly great for structures with heavy atoms in special positions, which cause certain classes of reflections to be systematically weak.

On the other hand, the presence of spurious reflections may lead to a cell which is larger than the correct one (or to no cell at all, because these reflections can not be made to fit). Allowance for some 'anomalous' reflections in the cell determination routine may help find an otherwise intractable cell, but it can also produce a subcell, rejecting genuine reflections in order to do so (Clegg, 1984b).

An effective method of avoiding these errors is the production of 'axial photographs' on the diffractometer, i.e. oscillation photographs taken about the three proposed cell axes or other lattice vectors. These can be checked for axial length and for symmetry. Alternatively, a systematic search can be made for reflections with fractional indices: this must be carried out for general reflections, and not just for projection and axial reflections,

which may be subject to additional systematic absences!
(b) Even if the three selected axes are of the correct lengths to characterise the lattice geometry, the cell may not conform to accepted conventions, nor even to the metric symmetry of the lattice. Errors of this type in published structures are regularly reported (e.g. Herbstein & Marsh, 1982). The use of cell reduction methods and comparison with tabulated standard forms, as incorporated into most commercial control software systems (Mighell & Rodgers, 1980) can itself lead to errors (Clegg, 1981b). Of the alternative methods proposed, the search for possible two-fold axes in the lattice (Le Page, 1982) is particularly effective, reliable, and comprehensible.
(c) The metric symmetry may be higher than the Laue symmetry, e.g. a monoclinic cell with $\beta = 90°$ is metrically orthorhombic (if not of even higher symmetry!). While the assumption of too low a Laue symmetry (assuming the correct symmetry is subsequently discovered) only leads in some cases to unnecessary extra data collection time, failure to recognise a 'too symmetric' unit cell may mean that only part of the unique intensity data is measured. The symmetry of axial photographs may help here. It is advisable to compare the intensities of supposedly equivalent reflections (but beware of the effects of absorption, etc!), and a routine for very rapid crude data collection (e.g. point measurement only) is useful. The use of complete sets of 'equivalent' reflections for cell refinement (see below) can also reveal an error in the assigned Laue group before it is too late.

2. Cell refinement

With high-precision intensity data, proper correction of systematic errors, and powerful computer programs, very low R factors can often be achieved. As a result, bond lengths are not infrequently quoted with esd's of around 2 in 10,000. Whether the precision and accuracy of the cell parameters justify this is questionable, unless special precautions are taken in their determination. Important factors are:-
(a) Diffractometer alignment; crystal centring and a stable mounting. Alignment should be checked regularly, using the same procedures as for normal reflection centring.
(b) Suitable reflections for cell refinement: 15-20 low-angle reflections do not give the required precision. Better cell angles are obtained from reflections well scattered in reciprocal space. The use of sets of equivalent reflections gives a good internal consistency check as well as confirming the assumed Laue symmetry.

(c) A suitable reflection centring method, e.g. a symmetric ω-scan profile can not be assumed for reflections with significantly but incompletely resolved α-doublet.
(d) Zero-point errors, particularly for 2θ. Determination of 2θ as the difference between two ω positions for centring of the reflection on each side of the direct beam eliminates the $2\theta_0$ error, as well as removing or minimising other errors due to crystal mis-centring, absorption, etc. (Bond, 1960).
(e) Proper refinement procedures. Although free 'triclinic' refinement may give the most effective orientation matrix for data collection, cell parameters for further calculations and publication should be derived with the correct symmetry constraints. Comparison of the constrained and unconstrained parameters and esd's can be very revealing concerning the difference between precision and accuracy!

3. Errors in data collection

Some potential blunders are obvious (or are they?): incorrect index limits leading to only a partial data set, a poor orientation matrix or wrongly set receiving aperture giving badly centred and possibly even 'clipped' reflections, overlapping reflections from structures with a long axis, etc.

More subtle are unwarranted assumptions in the estimation of intensities and their standard deviations, producing systematic bias in the data. This has been demonstrated for some methods (Tickle, 1975; Oatley & French, 1982), but may occur to a smaller degree in others. Application of the algorithm of Lehmann & Larsen (1974) in X-ray diffractometry leads to systematic errors, unless a correction is made (Blessing, Coppens & Becker, 1974; Grant & Gabe, 1978). Methods which assume a particular functional form for the reflection shape (e.g. Norrestam, 1972; Hanson, Watenpaugh, Sieker & Jensen, 1979; Oatley & French, 1982) can not be used when these conditions do not apply.

Any method attempting to exploit information from complete reflection profiles in order to improve estimates of I and $\sigma(I)$ should be subjected to extensive and rigorous consistency checks, to prevent unexpected circumstances from producing 'garbage' data (Diamond, 1969; Clegg, 1981a; Oatley & French, 1982).

The treatment of weak reflections is a bone of contention. They are not 'insignificant' (Arnberg, Hovmöller & Westman, 1979), and not to measure them at all after a rapid prescan may lead to many problems later. On the other hand, the use of long counting times, in an attempt to improve the precision, will be dreadfully wasteful for the weakest reflections. A compromise is necessary, and each data set must be treated as an individual case.

4. Wobbly crystals

Crystal movement during data collection is a recognised problem. It is usually monitored by the periodic measurement of a number of standard reflections (poor choice of these is another avoidable blunder!), which should be well distributed in reciprocal space. In commercial control systems, a routine is provided for re-establishing the orientation matrix and then continuing with data collection (Vandlen & Tulinksy, 1971). Reflections from a list are centred and the orientation matrix is refined without constraints (9 parameters). This effectively changes the unit cell parameters too. The time taken for this procedure can be considerable, and, if crystal movement continues during it, the exercise may be in vain.

An alternative is to assume an unchanged unit cell, and determine the (presumably small) rotation which much be applied to the orientation matrix to update it. This requires only 3 parameters, and can be achieved with as few as two reflections, though a larger number is advisable (Clegg, 1984a); for convenience, the same reflections can be used as standard reflections and for updating the orientation matrix.

5. Final remarks

Perhaps the most common blunder in data collection is to regard the whole process as a 'black box' and treat all samples more-or-less identically. This approach wastes considerable time and resources on crystals for which only gross stereochemical information is required, and may lead to disappointment when precise and detailed structural results are the aim. Although much fault can be found with diffractometer control systems, they are by no means always to blame for mistakes!

REFERENCES

Arnberg, L., Hovmöller, S. & Westman, S. (1979). Acta Cryst. A35, 497-499.

Blessing, R.H., Coppens, P. & Becker, P. (1974). J.Appl.Cryst. 7, 488-492.

Bond, W.L. (1960). Acta Cryst. 13, 814-818.

Clegg, W. (1981a). Acta Cryst. A37, 22-28.

Clegg, W. (1981b). Acta Cryst. A37, 913-915.

Clegg, W. (1984a). Acta Cryst. A40, in press.

Clegg, W. (1984b). J.Appl.Cryst. 17, in press.

Clegg, W. (1984c). Methods and Applications in Crystallographic Computing, ed. T.Ashida & S.Hall. Oxford: Univ. Press. In press.

Diamond, R. (1969). Acta Cryst. A25, 43-55.

Grant, D.F. & Gabe, E.J. (1978). J.Appl.Cryst. 11, 114-120.

Hanson, J.C., Watenpaugh, K.D., Sieker, L. & Jensen, L.H. (1979). Acta Cryst. A35, 616-621.

Herbstein, F.H. & Marsh, R.E. (1982). Acta Cryst. B38, 1051-1055.

Lehmann, M.S. & Larsen, F.K. (1974). Acta Cryst. A30, 580-584.

Le Page, Y. (1982). J.Appl.Cryst. 15, 255-259.

Mighell, A.D. & Rodgers, J.R. (1980). Acta Cryst. A36, 321-326.

Norrestam, R. (1972). Acta Chem.Scand. 26, 3226-3234.

Oatley, S. & French, S. (1982). Acta Cryst. A38, 537-549.

Tickle, I.J. (1975). Acta Cryst. B31, 329-331.

Vandlen, R.L. & Tulinsky, A. (1971). Acta Cryst. B27, 437-442.

EXPERIMENTAL DETERMINATION OF TRIPLET PHASES AND ENANTIOMORPHS.

By K. Hümmer and H. Billy

Institute of Angewandte Physik, Lehrstuhl Kristallographie, Loewenichstr. 22, University of Erlangen-Nürnberg, FRG

1. Introduction

It has been suggested for a long time that multiple beam X-Ray diffraction can be applied to determine the phase relationship of the waves involved (Lipscomb, 1949). In a so-called three-beam diffraction case three reciprocal lattice points (rlp) O,H,G lie simultaneously on or close to the Ewald sphere. Then, the amplitude of the wave propagated in a direction due to the reciprocal lattice vector (rlv)$\underline{h}$ results from the interference between the 'direct' wave diffracted at the netplanes of $\underline{h}$ and the detour excited wave (Renninger 'Umweg' wave) successively diffracted at the netplanes of $\underline{g}$ and $\underline{h}$-$\underline{g}$. Therefore, the amplitude of the resultant wave bears informations on the phase difference of the superposed waves:

$$\phi_{\Sigma} = (\phi(\underline{g}) + \phi(\underline{h}-\underline{g})) - \phi(\underline{h}) = \phi(-\underline{h}) + \phi(\underline{g}) + \phi(\underline{h}-\underline{g}),$$

which represents a structure invariant triplet phase relationship.

The change of intensity by this interference can be measured in a so-called Ψ-scan experiment monitoring the integrated intensity $I(\underline{h})$ as the crystal is rotated about the direction of $\underline{h}$ and scanned through a three-beam case. The profile of the Ψ-scan rocking curve depends on the triplet phase ϕ_{Σ} and it can be explained by a continuously turning on and off of the 'Umweg' wave amplitude and an additional phase shift $\Delta(\Psi)$ by π when the rlp G passes through the Ewald sphere (Hümmer & Billy, 1982). Bearing in mind that Bragg diffraction is a spatial resonance phenomenon this phase shift $\Delta(\Psi)$ is nothing else but the phase shift of every resonance phenomenon passing through the resonance.

Starting from simplified fundamental equations of the Dynamical Theory of three-beam X-ray diffraction and applying a perturbational approach, an effective structure factor is derived for a modified two-beam case which proves the validity of this interpretation. The interference between the directly diffracted wave and the 'Umweg' wave can be interpreted in terms of a phase-vector diagram which immediately outlines the general features of the Ψ-scan profiles even in the case of non-centrosymmetric structures.

In order to perform a well defined Ψ-scan a special six circle diffractometer was constructed. Experimental results will be reported for L-Asparagine with space group $P2_12_12_1$. The measured Ψ-scan profiles at least allow the determination of the quadrant of the triplet phase.

The experimental determination of structure invariant triplet phases also allows the fixing of the absolute configuration because the triplet phases of enantiomorphs differ in their sign and the Ψ-scan profiles for $\phi_\Sigma = \pm\pi/2$ are clearly distinguishable .

2. Three-beam Interference

In order to calculate the amplitudes of the X-ray wave fields in an ideal crystal one has to solve Maxwell's equations in a medium with a periodic complex dielectric susceptibility χ. This leads to the wave equation for the vector $\underline{D}$ of dielectric displacement

$$\frac{d^2}{dr^2}\underline{D} + 4\pi^2 k_o^2 \underline{D} = - \text{rot rot } \underline{P} \tag{1}$$

$$4\pi\underline{P} = \chi(\underline{r})\cdot\underline{D} \; ; \; |\underline{k}_o| = \frac{1}{\lambda_o} \; ,$$

which may be solved by the following relationships taking account of all the waves consistent with Bragg's law:

$$\underline{K}(\underline{h}) = \underline{K}(\underline{O}) + \underline{h} \; , \tag{2}$$

$$\underline{P} = \exp(2\pi i\nu t)\sum_{\underline{h}} \underline{P}(\underline{h})\exp(-2\pi i\;\underline{K}(\underline{h})\cdot\underline{r}) \tag{3}$$

$$\underline{D} = \exp(2\pi i\nu t)\sum_{\underline{h}} \underline{D}(\underline{h})\exp(-2\pi i\;\underline{K}(\underline{h})\cdot\underline{r}) \tag{4}$$

$$\chi(\underline{r}) = \sum_{\underline{h}} \chi(\underline{h})\exp(-2\pi i\;\underline{h}\cdot\underline{r}) \; ; \; \chi(\underline{h}) = -\Gamma F(\underline{h}) \tag{5}$$

$\underline{P}$ is the vector of dielectric polarization, $\underline{k}_o$ and $\underline{K}(\underline{h})$ are the wave vectors in the vacuum and inside the crystal, ν is the X-ray frequency, Γ is a small number of the order of 10^{-7} and

$F(\underline{h})$ stands for the structure factor.

The solutions of equ. (1) are called the fundamental equations of the Dynamical Theory:

$$\underline{D}(\underline{h}) = -\frac{K(\underline{h})^2}{K(\underline{h})^2 - K^2}\,\Gamma \sum_{g \neq h} F(\underline{h}-\underline{g})\,\underline{D}_h(\underline{g})\,;\quad |K| = |k_o|\,(1-\tfrac{1}{2}\Gamma F(\underline{O})) \qquad (6)$$

$\underline{D}_h(\underline{g})$ is the projection of $\underline{D}(\underline{g})$ on $\underline{D}(\underline{h})$; $|\underline{K}|$ is the radius of the Ewald sphere about the Lorentz point and it represents the wave vector in the one-beam case, i.e. only the rlp O lies on the sphere.

In his works about crystal optics of X-rays Ewald already pointed out that the dipole wave given by $\underline{P}$ travelling through the lattice is the "driving force" for the displacement field strength $\underline{D}(\underline{h})$ of a certain wave vector $\underline{K}(\underline{h})$. The resonance term $(K(\underline{h}^2)-K^2)^{-1}$ may be regarded as the efficiency of the crystal for converting a given amplitude of polarization $\underline{P}$ into the field amplitude $\underline{D}(\underline{h})$ (Ewald, 1965). If Bragg's diffraction condition $K^2(\underline{h}) = K^2$ is fulfilled this efficiency will be optimum. The amplitudes $\underline{D}(\underline{h})$ of waves, the corresponding rlp's of which lie outside $(K(\underline{h})^2 > K^2)$ respectively inside $(K(\underline{h})^2 < K^2)$ the Ewald sphere, undergo a change of their sign when crossing the sphere i.e. there is a phase shift by π going through the spatial resonance $K^2(\underline{h}) = K^2$ (Ewald, 1917).

For the interpretation of the three-beam interference we calculate the amplitudes of the wave fields near a three-beam case by a perturbational approach which is common in electron diffraction dynamical theory and often referred to by "Bethe potentials" (Bethe (1928), see also Juretschke (1982) and Høier & Marthinsen (1983)).

For simplicity we neglect the coupling terms due to different directions of polarization of the waves in the fundamental equations and so we are left with the following system of equations for a three-beam case with only one direction of polarization:

$$\begin{bmatrix} \frac{K(\underline{O})^2 - K^2}{K(\underline{O})^2} & \alpha_h^o F(-\underline{h}) & \alpha_g^o F(-\underline{g}) \\ \alpha_h^o F(\underline{h}) & \frac{K(\underline{h})^2 - K^2}{K(\underline{h})^2} & \alpha_g^h F(\underline{h}-\underline{g}) \\ \alpha_h^o F(\underline{g}) & \alpha_g^h F(\underline{g}-\underline{h}) & \frac{K(\underline{g})^2 - K^2}{K(\underline{g})^2} \end{bmatrix} \cdot \begin{bmatrix} D(\underline{O}) \\ D(\underline{h}) \\ D(\underline{g}) \end{bmatrix} = 0 \qquad (7)$$

α_g^h are geometrical coupling factors that are derived from the relations between $\underline{D}_h(\underline{g})$ and $\underline{D}(\underline{h})$.

If in a Ψ-scan experiment the third rlp G is far from the Ewald sphere then the field amplitude $D(\underline{g})$ is only small and it can be treated as a perturbation of the two-beam case O-H. The amplitude of the weak wave may be expressed approximately from (7) in terms of the amplitudes of the two strong waves $D(\underline{O})$ and $D(\underline{h})$. Then (7) can be reduced to give a modified two-beam case with an effective structure factor:

$$F_{eff} = \alpha_h^o F(\underline{h}) + \frac{K(\underline{g})^2}{K^2 - K(\underline{g})^2} \alpha_g^o \alpha_g^h F(\underline{g}) F(\underline{h}-\underline{g}) \qquad (8)$$

Then, the ratio $D(\underline{h})/D(\underline{O})$ is given by:

$$\frac{D(\underline{h})}{D(\underline{O})} = \frac{R(\underline{h})}{N} F_{eff} = \frac{R(\underline{h})}{N}(D_2(\underline{h}) + D_u(\underline{h},\underline{g})) \qquad (9)$$

with

$$R(\underline{n}) = \frac{K(\underline{n})^2}{K^2 - K(\underline{n})^2} \qquad \underline{n} = \underline{O},\underline{h},\underline{g}$$

$$D_2(\underline{h}) = \alpha_h^o F(\underline{h})\,; \; D_u(\underline{h},\underline{g}) = \frac{K(\underline{g})^2}{K^2 - K(\underline{g})^2} \alpha_g^o \alpha_g^h F(\underline{g}) F(\underline{h}-\underline{g})$$

$$N = 1 - \alpha_g^{h^2} F^2(\underline{h}-\underline{g}) R(\underline{g}) R(\underline{h})$$

In a first approximation N may be regarded as constant.

As can be seen from equ. (9) the amplitude $D(\underline{h})$ depends on i) the setting of the two-beam case O-H given by $R(\underline{h})$ and ii) on the interference between the directly excited wave (amplitude $D_2(\underline{h}) \sim F(\underline{h})$) and the Umweg wave, the amplitude $D_u(\underline{h},\underline{g})$ of which is given by the resonance term $R(\underline{g})$ and the product $F(\underline{g}) \cdot F(\underline{h}-\underline{g})$.

The phase relationship of both waves depends on two components: i) the triplet phase ϕ_Σ which is independent of the scanning angle Ψ, and ii) the resonance phase shift $\Delta(\Psi)$ by π.

This phase shift results from a change of the sign of $R(\underline{g})$ because it is $|K(\underline{g})| < |K|$ or $|K(\underline{g})| > |K|$ if the rlp G lies inside or outside the Ewald sphere respectively. Thus, the total phase $\phi(\Psi)$ depends on Ψ and is given by:

$$\phi(\Psi) = \phi_{\Sigma} + \Delta(\Psi) . \qquad (10)$$

In addition, the resonance term $R(\underline{g})$ governs the amplitude of the Umweg wave $D_u(\underline{h},\underline{g})$ as a function of the scanning angle Ψ. By rotation through the three-beam setting ($|K| = |K(\underline{g})|$) $D_u(\underline{h},\underline{g})$ is continuously switched on and off.

The superposition of the directly diffracted wave $D_2(\underline{h})$ and the Umweg wave $D_u(\underline{h},\underline{g})$ now can be outlined in a phase-vector diagram (Fig. 1), which essentially represents a graphical display of equation (9). The real axis defines the reference angle zero for all phases. We arbitrarily assign to $D_2(\underline{h})$ the phase angle zero, because only the phase difference between $D_2(\underline{h})$ and $D_u(\underline{h},\underline{g})$ is important. Then $D_2(\underline{h})$ is displayed by a vector on the real axis. If we assume a Lorentzian-type function for the spatial resonance term $R(\underline{g})$ - similar to the resonance term of forced vibrations of harmonic oscillators - scanning through the three-beam setting the vector terminal of the Umweg wave traces out a circle the real axis being tangent to it. This circle represents both the variations of the amplitude $D_u(\underline{h},\underline{g})$ and the resonance phase shift by π. The triplet phase ϕ_{Σ} is taken into account by rotating the tangent on the circle by ϕ_{Σ} with respect to the real axis. The sum of both vectors gives the resultant wave amplitude $D(\underline{h})$. If the Ψ-scan is performed so that G crosses the Ewald sphere from inside to outside then the vector of D_u runs along the circle in the sense as indicated in the figure. The dashed circle is a subsidary line. If the vector terminal of $D(\underline{h})$ lies outside of this circle then the resultant amplitude is larger, inside

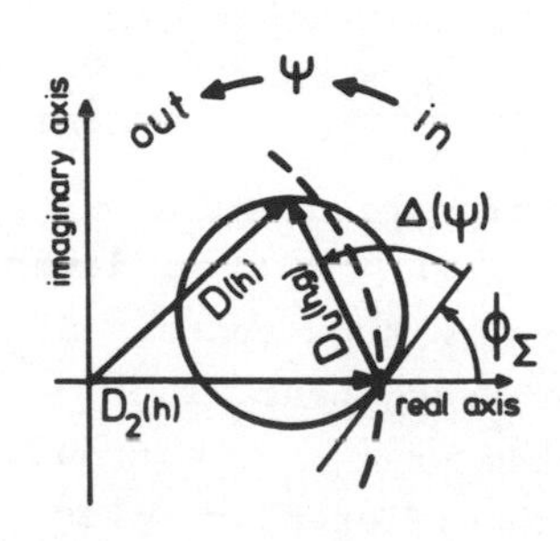

Fig.1. Phase-vector diagram of the three-beam interference.

it is smaller than the two-beam amplitude $D_2(\underline{h})$ giving a increased or decreased intensity respectively, compared to the two-beam intensity.

It is evident that this perturbational approach cannot describe exactly the fine structure of the Ψ-scan profile because it is not valid very close to the three-beam setting. For instance it does not take into account that the wave of $\underline{K}(\underline{h})$ couples back to the wave of $\underline{K}(\underline{g})$ via the rlv $\underline{g}-\underline{h}$. Furthermore, the denominator N (cf. equ.(9)) is not a constant near the three-beam position and this leads to a different behaviour in the case of a Bragg or Laue diffraction geometry.

3. Calculated Ψ-scan Profiles for Non-centrosymmetric Structures

In the following it will be shown that the general features of the Ψ-scan curves calculated on the basis of the full system of equations of the Dynamical Theory can be explained by the physical interpretation discussed above. The procedure for the calculations is outlined in detail by Hümmer & Billy (1982). As a non-centrosymmetric model structure we choose L-Asparagine: space group $P2_12_12_1$, lattice parameters a = 5,6 Å, b = 9,8 Å, c = 11,8 Å.

For the calculations the following parameters are used. Wavelength: $CuK\alpha_1$ = 1,540 Å; wavelength spread: $\Delta\lambda/\lambda = 3\cdot10^{-4}$; divergence of the primary beam: 2 minutes of arc; Ψ-scan sense: in-out; geometrical alignment: Bragg-Bragg or Bragg-Laue. The Ψ-scan rocking curves are calculated for the Bragg-reflection $\underline{h} = (2\bar{1}2)$ involved in various triplets with different triplet phases.

The results are shown in Fig. 2. In each case the corresponding phase-vector diagram plotted reveals the general features of the calculated curves. The indices indicated in the figures are the indices hkl of the Bragg reflections due to $\underline{h}$ and $\underline{g}$ in the three-beam case. However, the triplet phase is given by:

$$\phi_\Sigma = \phi(-\underline{h}) + \phi(\underline{g}) + \phi(\underline{h}-\underline{g}).$$

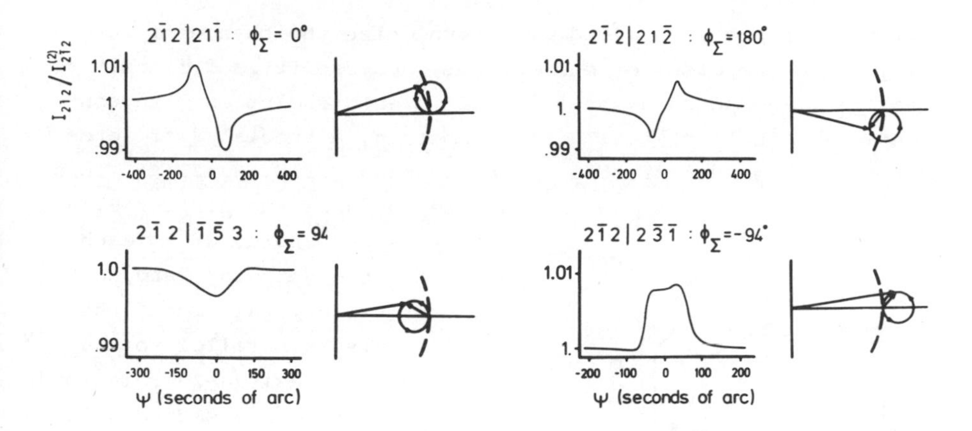

Fig. 2: Calculated Ψ-scan profiles for L-Asparagine with different triplet phases and the corresponding phase-vector diagrams.

4. Determination of Enantiomorphs

In non-centrosymmetric cases there are two enantiomorphic forms: the structure (S) with atom coordinates $\underline{r}_j$ and the inverse (I) with atom coordinates $\underline{r}'_j = -\underline{r}_j$. If the centre of symmetry is chosen as an origin for both forms the structure factors differ in the sign of their imaginary parts and their phases are related by $\phi_S(\underline{h}) = -\phi_I(\underline{h})$. Therefore, the triplet phase sums of S and I have opposite signs independent of the choice of the origin: $\phi_\Sigma(S) = -\phi_\Sigma(I)$. Hence, best selectors for S and I are triplet phases with $\phi_\Sigma = \pm\pi/2$. As both cases can be distinguished by means of the Ψ-scan profiles - near the three beam position one observes either a nearly symmetrical increase or decrease of the two-beam intensity - the absolute configuration can be fixed with only one measurement.

5. Experimental Results

5.1. Ψ-scan diffractometer

As can be seen from the calculated Ψ-scan curve the angular range in which the three beam interference can be observed is very small, about 3 minutes of arc. Moreover, when the three structure factors involved have equal size the change of intensity is in the order of one percent. Because it is difficult to realize a Ψ-scan experiment of sufficient resolution with conventional single crystal diffractometers, a special diffractometer has been constructed on the proposal of H. Burzlaff. This device contains two circles Θ, ν with axes perpendicular to each other for the detector, and four axes for the crystal. The first crystal axis is parallel to the first detector axis (ω-2Θ-relation); perpendicular to ω a second axis for Ψ-rotation is installed, this axis bears an Eulerian cradle with motions χ and φ. Thus an arbitrary scattering vector $\underline{h}$ can be aligned to the Ψ axis and a Ψ-scan is allowed moving only one circle (Fig. 3). During the measurement the detector lies in the horizontal plane through Ψ. Because of the two circles for the detector in the exact three-beam setting the second Bragg-reflection can be observed without moving the crystal.

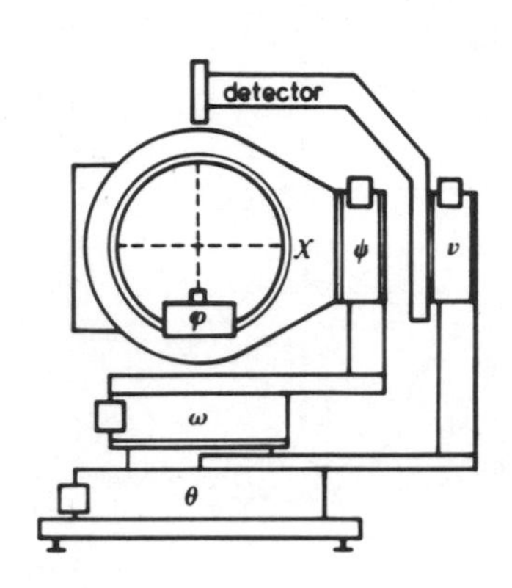

Fig. 3: Ψ-scan diffractometer

5.2. Measured Ψ-scan Profiles

As an example for a non-centrosymmetric structure we choose L-Asparagine: space group $P2_12_12_1$.

The parameters of the experimental set-up were as follows: radiation: $CuK\alpha_{1,2}$ (Cu anode with Ni filter); focus:0,15x0,8mm^2; distance focus to crystal: 0,6 m; divergency: 3 to 5 minutes of arc; measuring time for one point typical 200-300 sec; total time for the scan approximately 1-2h. All the profiles were plotted such that the positive direction means an in-out scan: for $\Psi<0$ and $\Psi>0$ the rlp G lies inside and outside the Ewald sphere respectively. Fig. 4 shows the four typical profiles already discussed in the theoretical part. It should be noticed that each profiles appears twice on the Ψ-scale because of $CuK\alpha_1$-$K\alpha_2$ splitting (Ψ=0 is the three-beam position for $CuK\alpha_1$). The splitting is clearly resolved for the two cases $\phi_\Sigma = 117^o$,

-113°, whereas the profiles for 0°, 180° can be interpreted as a superposition of two ideally asymmetric profiles in a distance of a few hundreds of seconds of arc. This gives a disturbance of the asymmetry of the profiles. From the measured profiles one immediately can deduce the quadrant of the triplet phase, i.e. whether the triplet phase is closer to 0°, 180° or $\pm 90^o$. The profiles for $\phi_\Sigma = 117^o$ and $\phi_\Sigma = -113^o$ confirm the absolute configuration of L-Asparagine.

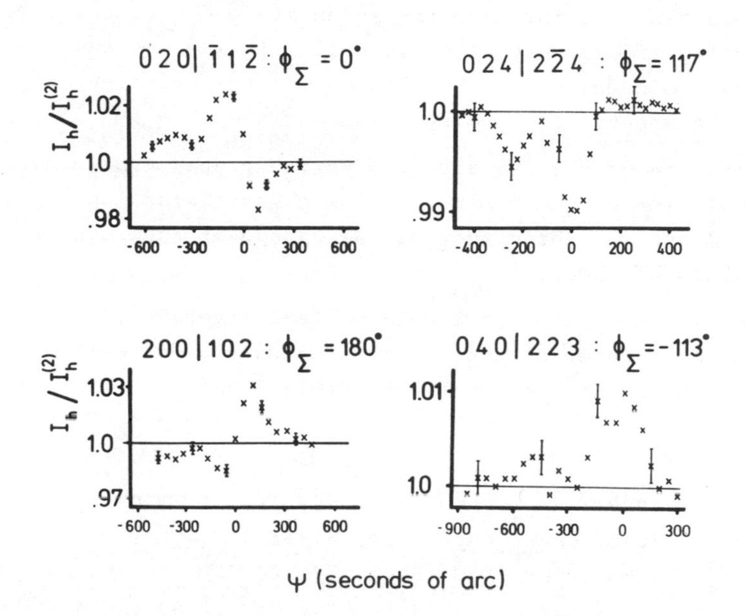

Fig. 4: Measured Ψ-scan profiles for L-Asparagine. Indicated are the triplet phase ϕ_Σ and the indices for the two Bragg reflections $\underline{h}$ and $\underline{g}$ in the three-beam case.

6. Discussion

If the amplitude of the 'Umweg' wave is too high there always results only an increase of the two-beam intensity. Thus, there is no essential difference between the Ψ-scan profiles of different triplet phases and the information about the triplet phase is lost. Therefore, the amplitude of the 'direct' wave and the 'Umweg' wave should be approximately of the same size, i.e., the products of the structure factors times the geometrical coupling terms α (the off-diagonal terms of equ. (7)) involved in a three-beam case should nearly be equal.

The different diffraction geometries - e.g. the Bragg case (back-reflection) and the Laue case (transmission) - show different kinds of behaviour. In transmission anomalous absorption effects may drastically change the typical Ψ-scan profiles. Therefore, the product of the linear absorption coefficient times the crystal thickness should be smaller than $\mu_o t < 0.5$. The Bragg case tends to increase the intensity close to the three-beam setting independent of ϕ_Σ (cf. Fig.2), whereas the Laue case tends to decrease it. This is due to the different sections through the sheets of the dispersion surface.

The divergence of the primary beam in the plane perpendicular to $\underline{h}$ should not exceed 5 minutes of arc, otherwise the profile is smeared out and the effect of the change of the intensity gets smaller and smaller.

It is not necessary to use ideal, perfect crystals to observe the three-beam interference. On the contrary, mosaic structure to a certain extent seems to be helpful.

References:

Bethe, H.A. (1928). Ann. Physik 87, 55-129

Ewald, P.P. (1917). Ann. Physik 54, 519-597

Ewald, P.P. (1965). Rev. Mod. Phys. 37, 46-56

Høier, R. & Marthinsen, K. (1983). Acta Cryst. A39, 854-860

Hümmer, K. & Billy, H. (1982). Acta Cryst. A38, 841-848

Juretschke, H. (1982). Physics Lett. 92A, 183-185

Lipscomb, W.N. (1949). Acta Cryst. 2, 193-194

COMPUTING ASPECTS OF POSITION-SENSITIVE DETECTORS

U.W. Arndt and D.J. Thomas

MRC Laboratory of Molecular Biology, Hills Road, Cambridge CB2 2QH, England

Abstract

Very fast and highly accurate diffractometry on large unit-cell single crystals is now possible using diffractometers equipped with position-sensitive, or area-, detectors. The success of these methods depends on the existence of powerful software packages which make considerable demands on the speed, memory size and data-storage capacity of the computer installation to which the diffractometer is interfaced.

Introduction

Many laboratories throughout the world now routinely collect X-ray diffraction data from crystals of biological macromolecules varying from small proteins and oligonucleotides up to complete viruses. The unit cell edges of these crystals extend from about 30Å to more than 300Å. There are two well-established methods to collect such data. The first is to use a single-counter diffractometer, which is accurate but slow since it can measure only one diffraction spot at a time. It is also wasteful of crystals because relatively few diffraction spots can be measured before the radiation damage to the crystal becomes too severe.

The second established method of data collection is the use of screenless X-ray cameras, mostly rotation cameras but also to some extent precession cameras. These X-ray cameras are efficient in their use of crystals because they can record all of the simultaneously diffracting spots concurrently, but the necessity to microdensitometer the film packs makes the method very time consuming.

There is consequently a widespread desire for an instrument which combines the collection efficiency of cameras with the accuracy of diffractometers without any loss in speed. It has been known widely for some time that a properly controlled electronic X-ray area detector could fulfil this role (see, for example, Arndt, 1978, 1983). There are two major classes of such detectors. One is based on multi-wire proportional-counter technology; the other is based on X-ray sensitive television cameras. Other detectors, for example one based on mosaics of superconducting bolometer beads, are possible, but have yet to be realised in practice. Useful surveys of X-ray area detectors will be found in the Proceedings of the International Conference on X-ray Detectors for Synchrotron Radiation (Bordas, Fourme and Koch, editors, 1982).

It is not the purpose of this paper to discuss the relative merits of the detectors, but rather to discuss the strategy of using them, and more particularly the computational difficulties in doing so. In fact, from the point of view of data collection strategy, the area detectors which have been realised are quite similar. They all have about 512 x 512 pixels, and therefore their spatial resolution is inferior to that of X-ray film. However the output signal from electronic detectors, be it a pure number representing the number of photons recorded in a given time or be it an analogue voltage characteristic of this number which is digitised in an analogue-to-digital converter, is approximately proportional to the incident X-ray intensity. When film is microdensitometered the quantity which is measured is the light transmitted through the exposed film; the X-ray intensity is proportional to the logarithm of this quantity. Since the sum of a series of logarithms is not equal to the logarithm of a sum it is necessary to sample the film on a much finer sampling raster (typically 2000 x 2000) than is needed with an area detector. Nevertheless, an important difference between film and area detector methods is that the coordinates of a diffraction spot in the detector plane can be recorded much more accurately with the former than with the latter. By way of compensation, area detectors permit a much better determination of the centre or peak of a reflexion in terms of rotational angle of the crystal spindle axis than do film methods.

Data Collection with Area Detector Diffractometers.

To collect a complete set of structure factors it is necessary to turn the crystal continuously about a given axis through a total angle which depends on the crystal symmetry. There is no particular advantage in not using a normal-beam method; the rotation axis is, therefore, usually made perpendicular to the incident beam. The diffraction pattern changes as the crystal turns and as the reciprocal lattice points pass through the Ewald Sphere. It is necessary to separate the X-ray intensity diffracted into a given spot not only from that in neighbouring reflexions but also from the background scattering. It must be emphasized that this X-ray background in any diffraction pattern is proportional to the total volume of reciprocal space illuminated during the 'exposure'. This volume necessarily increases as we go from zero-dimensional data collection (single-counter diffractometer) and two-dimensional methods (moving film photography with layer-line screen) to three-dimensional methods (screenless rotation photography).

Two possible strategies of data collection come to mind immediately. In the first, the area detector diffractometer mimics a rotation camera: the patterns produced by the turning crystal are recorded as a contiguous series of small-rotation-angle "photographs", rather like a cine film. In rotation photography proper the angular range corresponding to a given exposure is chosen so as to record the maximum number of spots without spatial overlap, and thus to minimize the number of partially recorded reflexions on the film. This angular range is frequently much greater than the reflecting range of any one reflection. Accordingly, the signal-to-background ratio is not optimised because the criterion that the volume of reciprocal space illuminated should be minimised is not satisfied. When an area detector diffractometer is used in the "camera mode" the angular range for each exposure can be made several times smaller than the reflecting ranges of individual spots and the signal-to-background ratio can be greatly improved by three-dimensional profile fitting. With a stable area detector, of course, there is no problem akin to that of inter-film scaling in rotation photography which makes the measurement of reflexions partially recorded on more than one film less accurate. The considerable data storage problems in this camera mode of data collection will be discussed below.

The second method of data collection, can be described as the "diffractometer mode". It is an attempt to measure the diffraction spots where and when they occur without incorporating too many background counts. The computational and data storage demands of the two methods are very different and will be treated separately, though much greater emphasis will be given to the more advanced and complex diffractometer mode.

The Camera Mode

In the camera mode at its simplest, one needs to dump a large number of high-resolution images to a mass-storage medium. Intrinsically easy, this is rendered difficult only by the speed with which it needs to be done and by the colossal volume of a complete set of data. To gain an idea of the size of the problem, consider a typical modern detector with a resolution of 512 x 512 pixels. In an integrating television type of chamber with a 16-bit word output this makes ½ Mbyte per image; a multiwire chamber would need a similar storage capacity. To collect data of the highest accuracy we would wish to collect about ten small-angle rotation photographs per degree of arc, spread over a total rotation range of over 90° for a crystal of low symmetry, plus perhaps 10° or more to fill in the cusp of the torus of revolution. There are thus at least 1,000 images each of ½ Mbyte resulting in a total data volume of 500 Mbytes. This is not totally unreasonable but it does present problems, which we shall now examine.

The rotational speed of the crystal during data collection is about the same as that which would be used with a single-counter diffractometer, scaled up or down according to the volume of the specimen crystal which, in turn is dictated to some extent by the physical dimensions of the detector. For a multi-wire proportional chamber the ideal volume is about the same as would be used in diffractometry; for a television detector with a detector only about 50 mm square the specimen crystal would be, by preference, several times smaller. A reasonable speed for a 100Å unit cell crystal on a television diffractometer equipped with a sealed X-ray tube is about 0.2 degrees minute^{-1}, that is, each individual one-tenth degree image will be recorded in 30 sec. Accordingly the data rate is 17 kbytes sec^{-1}. Even with rotating-anode tubes (but not with synchrotron radiation sources), the transfer rates cause no difficulties, but the data volume does. A 2 Mbyte removable cartridge disk can only hold 4 images and would be useless for data storage; a 250 Mbyte Winchester disk could be used, but would need to be unloaded half-way through and at the end of the run, yielding a total of 12½ reels of 1600 bpi magnetic tapes each holding 40 Mbytes. If the data were recorded directly on to tape the reels would have to be changed every 40 minutes.

The Diffractometer Mode

In the camera mode a complete uncontracted data set is collected and subsequently analysed off-line. The second method of data collection, which we call the diffractometer mode, works by predicting when, where and for how long each diffraction spot will appear as the crystal is rotated at a uniform rate about a chosen axis. The motion of the crystal and its lattice geometry determines the sequence in which diffraction spots appear and in which they must be measured; the measuring procedure is thus quite different from that in conventional single-counter diffractometry where spots are measured one at a time and in any convenient order.

The algorithms used for predicting diffraction spot positions are most conveniently based on the Ewald Sphere construction. The difficulties in the computation are in the prediction of the sequence in which reciprocal lattice points cut the sphere; this is a problem which has no analytical solution and which must, therefore, be solved algorithmically.

The first attempts of solving the prediction problem consisted of permuting all values of the spot indices, (h,k,l), which were of interest and sorting this enormous list into order of the angle of the rotation axis at which each spot would occur. This approach has several disadvantages; one is that it is very time-consuming to create and sort such lists; another is that it demands that the geometry of the experiment be known exactly before it is begun, and that it is unchanging.

To circumvent these problems, and to provide more versatility, a new algorthm was developed to be run in real time, concurrently with the diffraction experiment, simulating the experiment as closely as possible. This algorithm has been called GENREF; it can also be used in film work if desired. It is important, as in all diffraction experiments, that accurate predictions be made also of the range of angles over which a spot diffracts, the so-called diffraction width. The conventional equations for the width were thought to be too cumbersome and not properly representative, so a new geometric analysis has been made, which led to a versatile and elegant analytic form well suited to fast computation.

Three important advantages accrue once an accurate and fast real-time prediction and measurement of diffraction spots has been achieved. Firstly, the volume of information which has to be recorded is greatly reduced, as measurements are made only in the immediate vicinity of reciprocal lattice points instead of throughout reciprocal space. Secondly, the more accurate our knowledge of the expected three-dimensional shape of the diffraction region in reciprocal space, the better is the signal-to-background ratio which we can achieve. Finally, we can monitor the progress of the experiment continuously, and, if necessary, adapt to any changing circumstances such as, for example, changes in the orientation of the crystal due to slow slippage.

This adaptation process consists of a refinement of many parameters, such as the orientation matrix and unit cell parameters, as the experiment proceeds. It poses quite difficult analytical as well as computational problems. The latter arise because the large quantity of data produced does not appear in a form which is particularly well suited to analysis, and must therefore, be at least partially collated in the computer memory.

The complete set of programs, comprising those to predict and measure the diffraction spots, to output the data, to monitor the course of the experiment and to make corrections as necessary, is quite complex. (These programs will be discussed in more detail in the companion paper by Thomas (1984)). The rate at which spots come into the diffracting position is far from constant and necessitates the use of a certain amount of internal buffering between various parts of the programs. A system in which the experiment is monitored and in which the predictions are corrected is a servo system; its stability and accuracy of response are reduced as the length of any buffers increases. A certain amount of judgment has to be exercised in choosing the correct length of these buffers, and they have to be adjusted continuously during data-collection by a scheduler designed to maintain the stability of the whole system.

At the same time, the computer to which the diffractometer is interfaced must service interrupt requests and output data to disk. With an area-detector diffractometer both of these tasks must be performed at a relatively high rate.

Experimental Corrections

All area detectors pose their own special problems. Firstly, as we have seen, their spatial resolution and accuracy are inferior to those of X-ray film, nor can diffraction angles be determined with the same precision as on a single-counter diffractometer with its accurate 2θ arm. The all- important initial determination of the orientation matrix and of the unit cell parameters must, therefore, use new algorithms which depend almost exclusively on the location of the centres of the diffraction spots in ϕ which can be very accurate.

Secondly, electronic detectors always produce slight distortions of the image which must be corrected. For television detectors these distortions are quite large but they are fairly regular and slowly varying, so that they can be corrected easily by low-order polynomial expansions. For multi-wire proportional chambers the distortions tend to be on a very small scale which is related to the periodicity of the read-out wires. When an area-detector is used in a diffractometer mode any image distortion sufficiently severe to endanger data-collection must be corrected in "real-time". This turns out to be simple but constitutes another demand on computing time.

Thirdly, all area-detectors suffer from variations of response across their area. In general these variations tend to occur on all spatial scales to some extent or another. Thus television detectors have small random local variations superimposed on a much more gradually falling response at the edges and especially at the corners. A typical multiwire device on the other hand, tends to have more strongly periodic local variations of sensitivity with no particular fall-off at the edges, though there is often a marked dependance on the angle of incidence of the detected X-rays. Regardless of the type of detector, and of the mode of operation, corrections for non-uniformity of response can be made either concurrently with the diffraction experiment, or afterwards. In diffractometer mode, though, the real-time computations are already so complex that all non-essential corrections are best left till later.

Computer Requirements

The demands on the data storage devices and on the computer in the diffractometer mode are rather different from those in the camera mode and depend on whether real-time refinement is carried out or not. If profile analysis is done off-line the amount of information to be carried for each reflexion is about one block (512 bytes). Thus for either a 100Å unit cell crystal measured to 2.5Å resolution or a 200Å unit cell crystal to 5Å the total number of spots is 100,000 and the complete output is about 50 Mbytes, a reduction by a factor of 10 as compared with the output data volume in the camera mode. If on-line profile analysis and refinement are carried out the total output is reduced by another factor of about 25 to a total of 2 Mbytes.

While the demands on output and storage are relatively light in the diffractometer mode the demands on the computer are severe, both as regards computing speed and core-store. Our prediction programs, written in FORTRAN, are capable of predicting about 100, 30 and 3 reflections sec^{-1}, respectively, when running on a VAX 11/780, VAX 11/730 and PDP 11/24 computer. We have seen above that a data collection run on a 100Å unit-cell crystal to 2.5Å resolution with a sealed X-ray tube source lasts about 30,000 sec so that the mean number of reflections per second is about 3. The peak rate, when a densely populated reciprocal lattice plane is tangential to the Ewald Sphere, may be very much greater than the mean. At present, therefore, even without profile analysis and refinement, the use of a PDP 11/24 or PDP 11/23+ computer may well be the limiting factor in the data collection rate.

The size of the computer decides the method of data collection. Our programs, excluding profile analysis and refinement programs, run to about 30k words of machine code when overlaid in a PDP-11 and the total store required is about 220k words, including about 100k words of data space, so that a computer with 512 kbytes of memory is required. On-line profile analysis and refinement are still not possible with a computer of this size as they require at least 1 Mbyte of readily accessible memory. Users of existing ENRAF-NONIUS television diffractometers are planning to connect these instruments to small VAX-type computers as soon as they become available.

Data collected so far with area-detector diffractometers have shown that the claims which have been made in the past for precision and speed were completely justified but can be met fully only when they are connected to powerful data acquisition computers.

If use is to be made of the data collection speeds which are, in principle, possible with rotating-anode X-ray generators and with synchrotron radiation sources existing programs may have to be speeded up. It may become necessary to use front-end multiple-processor computers to speed up program execution even further to make possible complete data collection from macro-molecular crystals in periods of a few minutes which are now physically possible.

References

Arndt, U.W. (1978). in Computing in Crystallography eds. H. Schenk, R. Olthof-Hazekamp, H. Van Kongingsveld and G.C. Bassi, Delft University Press.

Arndt, U.W. (1983). I.U.Cr. Summer School on Computing, Kyoto, 1983, Proceedings to be published by Oxford University Press, eds. T. Ashida and S.R. Hall.

Bordas, J., Fourme R. and Koch, M.H.J. editors. X-ray Detectors for Synchrotron Radiation. Nucl. Instrum. and Meth. 201, 1-279, 1982.

Thomas, D.J. (1984). These proceedings, p.

COMPUTING FOR THE ENRAF-NONIUS FAST-SYSTEM

By D.J. Thomas

M.R.C. Laboratory of Molecular Biology
Hills Road, Cambridge, CB2 2QH, England

Abstract

Strategies of data-collection with area-detector diffractometers are discussed, and the practical implementation of these strategies on a commercially available system is described.

Introduction to the ENRAF-NONIUS FAST-diffractometer

This paper discusses the computing techniques developed for the ENRAF-NONIUS FAST-system. More general aspects of the control of position-sensitive detectors are discussed in the companion paper by Arndt & Thomas (1984).

The FAST-system is a commercial development by ENRAF-NONIUS, Delft, of the television-type of X-ray area-diffractometer developed in Cambridge (U.K.) by Arndt and others (1968, 1969, 1973, 1975, 1977, 1982). Detailed information about the system is contained in the handbook to the instrument.

The strategies of data-collection

All currently available area-detector diffractometers are used in a mode resembling conventional rotation photography to a greater or lesser extent. This is because the same arguments about the proper tactics of collecting diffraction data efficiently on film which were used to justify the development of the screenless rotation camera apply equally when using an electronic area-detector. This means that they also suffer from the "cusp problem" which relates to the region of reciprocal space which cannot be measured by a single rotation of the crystal about one axis. The problem is overcome with the conventional rotation camera by rotating about another axis, and this method can also be used with electronic area-detectors. All multiwire proportional counter systems currently work in the so-called "camera mode" (Arndt & Thomas (1984)), taking long series of small-angle rotation pictures. The ENRAF-NONIUS FAST television based system has an alternative as well, the so called "diffractometer mode" of usage (idem), in which the crystal rotates continuously, and the spots are observed individually and concurrently as and when they appear, never being presented in the form of complete images. This method is known as the continuous rotation strategy, and can be accomplished using the ENRAF-NONIUS Phase-III FAST software. The strategy is capable of variation: if for example, the crystal is rotated sufficiently slowly that all diffraction spots are measured to an adequate statistical precision, then it is called the slow continuous rotation strategy; but if, on the other hand, the crystal is rotated faster so that no single measurement is of the required statistical precision, then it is called the fast continuous rotation strategy. The spread of systematic errors caused by radiation damage is different in the two strategies, and each will be appropriate in different circumstances. The aim of the fast rotation strategy is somewhat unconventional, in that it is intended to provide a complete torus of data from each crystal, and different crystals are supposed to be rotated about different axes. This enables the cusp of the torus of revolution to be filled in, and spreads the systematic errors of data collection more uniformly. More complex strategies are possible, and we hope to develop these later.

Introduction to Practical Implementation

When the ENRAF-NONIUS FAST-diffractometer was first being developed, it was planned to release the program packages in three phases as the successively more advanced versions became available.

Phase I software, which was released with the first instruments, treated the instrument as if it were a photographic rotation camera. That is, it took a series of small-angle rotation photographs and dissected them later. The crudity of this method was recognised from the beginning, and it inevitably limited the performance of the instrument.

The Phase II software was designed along the lines set out in Arndt & Wonacott (1977). In essence, it was to have precomputed an ordered list of reflexions and then to have asked the diffractometer to examine just that small region of reciprocal space around each spot as it appeared. This method has the enormous advantage of avoiding the incorporation of large volumes of background intensity into the data. However, it also has the disadvantage that the precomputation is liable to take as long as the diffraction experiment itself, thus halving the usable speed of the instrument. As it happens, Phase II was overtaken by Phase III and was never developed.

Phase III was designed to have the same Phase II type of method of a "measurement window" around each spot in reciprocal space which is so fundamental to the efficient measurement of the diffraction pattern, but the predicted diffraction pattern evolves in step with the real one. It is almost self-evident that there is no faster way to measure a diffraction pattern than to turn the crystal constantly, always measuring every spot which appears, so Phase III can be described as an optimal method. Yet another advantage accrues from this strategy, as well, and this is that since the occurrence of each diffraction spot is predicted only a few seconds before it actually appears, it is possible to take into account any small changes in the experimental conditions as they occur. For example, should the unit cell of the crystal slowly skew, as it does with _actin_, or should the angular width of spots slowly increase with radiation damage, or should the crystal slip slowly in the capillary then these and many other slowly changing disturbances can be accommodated and data-collection can continue unimpeded. The ability of the Phase III programs to adapt to the prevailing conditions means that it is even possible to start a data--collection run before the orientation and cell matrices are known to what would conventionally be regarded as a sufficient accuracy.

The suite of programs comprising Phase III naturally include some parts in common with the earlier phases. These include general utility facilities and some specific crystallographic programs such as SEARCH and INDEX which find diffraction spots in a still or small angle oscillation photograph and index them, thus producing an oriented lattice matrix for the crystal. These programs are derived from various pre-existing sources and some have been newly developed by ENRAF-NONIUS. These parts of Phase III are described in the handbook to the instrument.

In addition to these well-established programs, Phase III includes a substantial body of completely new software written (by the author) especially for high-speed area-detector diffractometers in general and the FAST-system in particular. This new software is composed of six distinct but mutually dependent subprograms. Their names and brief descriptions of their functions are as follows:

1) PREGEN (prepare for reflexion generation)
2) INIGEN (initialise reflexion generation)
3) GENREF (generate reflexions)
4) MESREF (measure reflexions)
5) REFINE (refine our knowledge of experimental parameters)
6) UPDATE (update our estimates of the experimental parameters).

Of these, PREGEN and INIGEN must have been run before data-collection can start and can be regarded as a vestigial but admittedly complex precomputation stage in the entire data collection sequence. MESREF must run continuously whilst data are being collected, and GENREF must run almost continually.

These first subprograms are sufficient to collect data using the continuous rotation strategy in the event that both the crystal and the beam are sufficiently accurately described at the beginning of the run, and neither are subject to any significant change thereafter.

REFINE and UPDATE are the gilding on the lily. REFINE itself is in a sense a diagnostic routine which collates diffraction data in such a way that it is always possible to determine a good estimate of the current crystal and beam properties. It can be run with GENREF and MESREF but without UPDATE if so desired. UPDATE itself makes use of the information collated by REFINE and actually makes any necessary alterations to the parameters which describe the geometries of the diffraction experiment. UPDATE and REFINE communicate in such a way that they can be run asynchronously and always operate in a statistically optimum way.

The major Phase III algorithms

PREGEN

The first of the Phase-III algorithms proper is called PREGEN. Its job is to create a look-up table called "the map" which guides the main prediction algorithm. It describes a method of reindexing reciprocal lattice points in a way which makes the prediction of diffraction patterns easier and efficient. Every diffraction spot in GENREF is predicted from one which is known already by translating it into the region which will soon be swept by the Ewald sphere. This corresponds to a rotational shift about the axis. The set of points predicted one from another lie on a line known as a prediction path. The simplest way to organise the prediction is if all prediction paths imitate concentric circles centred on the axis. Reciprocal lattice points lie on a regular lattice, though, so the paths actually used must be polygonal. It is also required that every diffraction spot be predicted exactly once, not several times or not at all, so the polygonal paths must be chosen properly in relation to their neighbours, and must not overlap each other or have gaps between them. In order to satisfy this condition, it is generally necessary that many variations of the basic shape of the paths be allowed. If all of the paths were genuinely different, this would pose computational difficulties, but it turns out that a combinatorial method exists which will generate them automatically. It works by considering a set of supercells of the reciprocal lattice (that is cells of a sublattice) and finds a mapping between their contents. Every reciprocal lattice point occupies a characteristic position in the supercells, and so long as this is known, it is an easy matter to look up which mapping to use to create the next diffraction spot along the prediction path. It is PREGEN's role to establish which supercells of the lattice to use, to find the best mapping between them, and to code the results in a form suitable for INIGEN and GENREF to read.

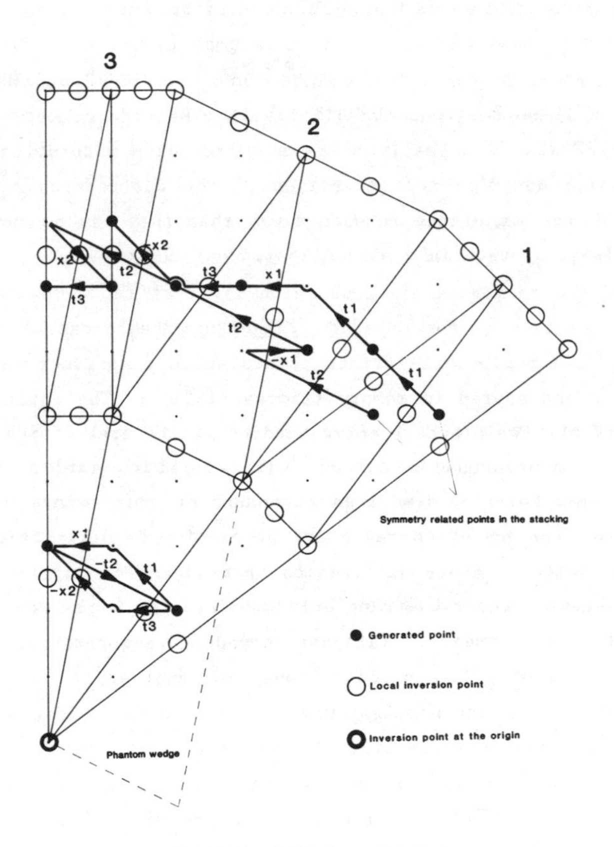

The prediction of diffraction spot positions

PREGEN does not in any way determine the exact course of the prediction algorithms, it is strictly a program which specifies the tactics which those later programs will use to predict the diffraction pattern. It does not, therefore, need to know anything about the geometry of the X-ray source, the diffractometer or the detector, nor does it need to know about the mosaicity properties of the crystal. It simply needs approximate estimates of the unit cell matrix, the axis that the crystal will be rotated about, and an upper limit on the area of the diffraction pattern to be observed. Because of this, it need be run only once for any series of experiments using crystals of a given unit cell rotated about nominally the same axis. Its output map will be equally valid for crystals whose cell parameters differ by a few percent, or whose interaxial angles differ by a degree or so, and a tolerance of several degrees is allowable on the rotation axis actually used.

Even the simplest versions of PREGEN are very complex, and the code runs to many thousands of lines of FORTRAN, but despite this it runs quickly and reliably.

INIGEN

INIGEN is the algorithm which is used to set up the start of data--collection proper. It does this by enumerating explicitly the contents of supercells of the reciprocal lattice which lie near to the Ewald sphere at the starting position of the crystal. These supercells can be described as "filled" once the reciprocal lattice points which comprise their contents are enumerated and stored in memory (Thomas (1981)). The entire region of the surface of the Ewald sphere corresponding to the area of the diffraction pattern to be measured must be covered with filled supercells. The supercells themselves, being basically brick-shaped, cannot form a particularly good representation of the surface of the Ewald sphere, so the initial distribution is smoothed out and made to drape over the Ewald sphere by an efficient analytic transformation. INIGEN finally sorts the points comprising the initial diffraction pattern in the manner required by GENREF.

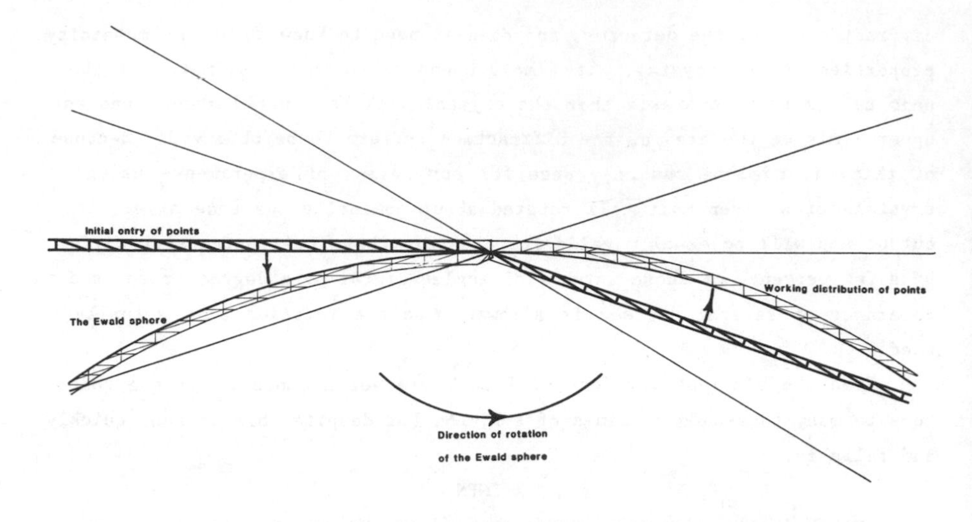

Starting the generation algorithm

Because of its role in initialising data-collection, INIGEN requires an extensive set of input data, including accurate terms describing the crystal's unit cell, orientation and mosaicity, the beam alignment, cross-fire and wavelength distribution, the required area of the diffraction pattern to be observed and the angle of the crystal at which data-collection should start. It does not, however, require any information about the geometry of the detector.

GENREF

GENREF, the principal "run-time" algorithm, predicts the continuously changing diffraction pattern falling on the detector. It is a subroutine, and it predicts one new diffraction spot each time it is called. Algorithmically, it usually does no more than move a reciprocal lattice pointer one step along a lattice line and sort the new pointer position in order of its diffraction angle. However, there are conditions when it does more than this: one is when it becomes appropriate to predict new points by jumping along a different lattice line as the crystal turns; the other is when it predicts points lying near to the rotation axis, in which case a related recursive procedure is adopted. Its operations are predetermined by the map produced by PREGEN, but the order in which they are done depends on the exact diffraction geometry as input to INIGEN.

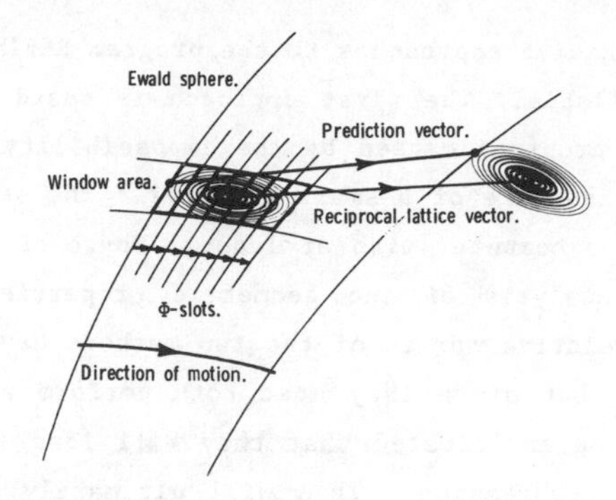

The prediction of diffraction spot positions from those already known to be on the Ewald sphere.

MESREF

MESREF accepts the preliminary predictions of diffraction spots from GENREF and performs all of the operations necessary to measure them. First it improves the accuracy of the calculations of diffraction angle and width using the best current estimates of the experimental conditions and it determines the position of the spots on the detector, taking into account any known image distortions. Then, taking account of the crystallographer's requirements and the need to prevent arithmetic overflows in the hardware, it decides into how many angular slots each spot should be split in order to measure it correctly. It then determines the correct sequence for the measurements of the spots and directs the diffractometer to do so.

MESREF needs to be told about the geometry of the detector and the angle through which the crystal is to be turned in addition to the information already input to PREGEN and INIGEN. Its interface to the diffractometer is designed not to be machine-specific.

REFINE and UPDATE

REFINE and UPDATE are very closely related coroutines responsible respectively for monitoring the true geometry of the diffraction pattern and for making any necessary changes to the parameters controlling the prediction and measurement of the spots. The geometrical analysis used is rather unusual, in that it relies almost exclusively on the observed diffraction angles of the reflexions. This is done because the angular positions of diffraction spots can be determined for more accurately than their positions in the plane of the detector. The angular widths of spots are also used, which is why it is so important to have a simple but accurate analytic approximation for them.

There are two alternative approaches to the program REFINE depending on the computing power available. The first approach is based on a special trick to circumvent the problems caused by the impossibility of collating the diffraction data in the core of a small computer. The second approach, which can be run only in a computer with at least 1 Mbyte of core, is based on a full statistical analysis of the geometric properties of the diffraction data. The relative merits of the two methods have not yet been examined experimentally, but since they must both perform a rudimentary profile analysis it can be anticipated that they will lead to a reliable method of run-time data-compression. This will ultimately render area--detector diffractometry by far the most convenient method of data--collection from macromolecular crystals.

Conclusion

It may seem curious at first that so many new programs were needed just to measure diffraction patterns efficiently in "real-time". This is at least partly because they were designed to be so flexible and versatile from the outset, and great emphasis has been laid on making them "user-friendly". Despite their size and complexity the Phase-III programs have been tested extensively and are known to run reliably. They will be released to users in the near future, and it is to be hoped that excellent results will be obtained with them.

References and Bibliography

Arndt, U.W. (1968) Acta. Cryst. B24, 1355.

Arndt, U.W. (1969) Acta. Cryst. A25, 161.

Arndt, U.W., Champness, J.N., Phizackerly, R.P. & Wonacott, A.J. (1973) J. Appl. Cryst. 6, 457.

Arndt, U.W. & Gilmore, D.J. (1975) Adv. Electron. Electron Phys. 40B, 913.

Arndt, U.W. & Wonacott, A.J. (1977) The Rotation Method in Crystallography (North-Holland, Amsterdam)

Arndt, U.W. & Thomas, D.J. (1982) Nuc. Inst. & Methods 201, 21.

Arndt, U.W. & Thomas, D.J. (1984) These proceedings. p.

Thomas, D.J. (1981) Acta.Cryst. A37, 553.

Thomas, D.J. (1982) Nucl. Inst. & Methods 201, 27.

Thomas, D.J. (1983) Ph.D. Thesis, University of Cambridge.

Experiences with a novel "fast transmission powder diffractometry" using a Stoe made curved position sensitive proportional counter with an angular range of Delta 2theta = 50 degrees

by E.R. WOELFEL
STOE APPLICATION LABORATORY D-6100 DARMSTADT

The fast transmission diffractometry which is based on a focused X-ray beam path in conjunction with the PSD is continuously used since about 3 years in our laboratory. The 1982 state has been described in JAC (1983) 16, 341-348. Since then the magnetic field within the PSD and the electronic components have been improved.

After a discussion of the X-ray beam path and the working principle of the PSD some characteristic features of the novel diffractometry will be discussed in this paper.

I. The X-ray beam path

The beam path of the focused monochromatic X-ray beam is shown in Fig. 1.

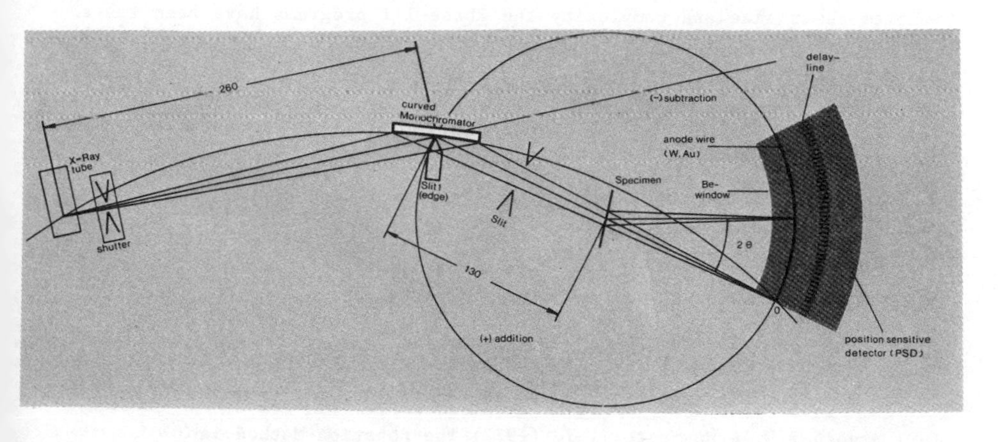

Fig. 1:
Path of the monochromatic convergent K(alpha1) X-ray beam

The curved Germanium monochromator which is mounted on the circumference of the diffractometer supplies a homogenuous convergent K alpha 1 beam, the cross section of which is 8 x 2 mm at the specimen. This beam is focused at the point 0 on the diffractometer circle (2theta = 0 degree) and provides X-ray pattern of excellent angular resolution and sufficient intensity if flat specimen or capillaries are used and if some precautions are observed in specimen preparation (see section III below). Thus most complicated powder patterns can be decomposed and indexed without any difficulties.

For practical applications this beam path has the following advantages:

- Easy adjustment of the convergent beam (K alpha 1-line) over a centred capillary. (The diffractometer is rotatable around the vertical axis of the monochromator). Thus the wavelength can easily be changed.
- Correct adjustment of samples into the diffractometer centre with a high resolution microscope.
- Step-scan adjustment of 2theta = 0 (to +/- 0.001 degree routinely).
- Flexible arrangement of special collimators (pin hole for single crystals and fibres, narrow slits for capillaries etc.).
- The absolute peak position accuracy is better than Delta 2theta = +/- 0.005 degree for routine scans.

One of the most important features of the beam path is the possibility of using the Stoe made PSD, the anode wire of which coincides with the focusing circle of Fig. 1.

II. The working principle of the Stoe PSD

The PSD which was developed and made by Stoe is shown in Fig. 2. It is a proportional counter which is filled with a Ar/CH4 mixture at 5 atm, absorbing approximately 35% of Cu K alpha and 80% of Co K alpha radiation. The Argon is ionized by the X-ray photons and the electron avalanches reach the positive anode wire (W,Au plated), the electric charges are transferred to the delay line and the delay line pulses are amplified and further processed by constant fraction discriminators to produce precisely timed pulses which control the output of a time to amplitude converter (TAC).

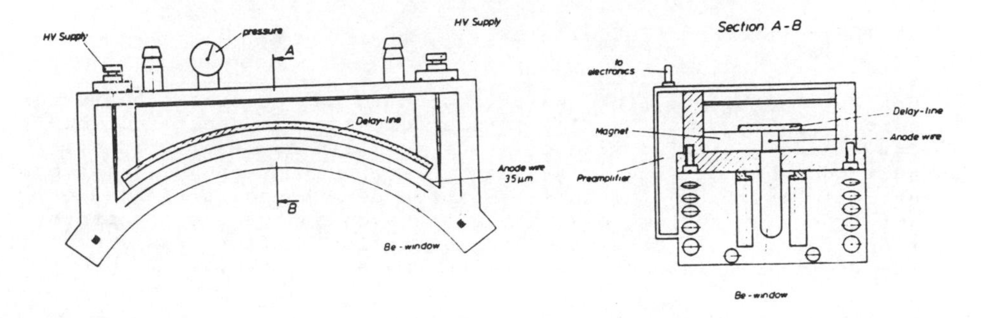

Fig. 2: The Stoe position sensitive proportional counter (PSD)

The pulse from one end of the delay line is used as a start pulse and the other, which is sufficiently delayed to ensure it arrives after the start pulse, is used as the stop pulse. The analog signals are then digitized and the events are stored in the memory of the PSD interface which has been designed by Stoe. The fast data transfer to the computer is made after the completion of measurement via a parallel interface and takes only a few seconds.

A special feature to which we refer later is the display of the analogue signals of the TAC on an oscilloscope. This display makes it possible to observe even small changes of diffraction pattern which occur within time intervals of 0.1 - 0.2 sec.

III. Some features of the novel fast transmission powder diffractometry

Some characteristic features of the new diffractometry which became evident during its continuous use in our application laboratory will be described in this section.

a) The role of the TAC display

On the TAC display the analoque signals can be continuously observed which enables one, to characterize the specimen. Thus a few larqe particles within a specimen can be easily identified, homogenuous particle size distributions can be observed, optimal specimen thicknesses can be determined, preferred orientation is easily detectable and even fast phase transformations of sinqle crystals can be followed after rotation of a large particle into its reflecting Bragg position.

b) Exposure time for PSD scans

The exposure time for routine PSD transmission diagrams is 2 - 3 min, whereby about 10 **6 counts are accumulated over the total pattern if well crystallized samples with low background are studied. From highly disordered samples as polymers etc. up to 3 x 10 **6 counts may be obtained within about 3 min.

Only in exceptional cases exposure times up to 5 min are required. For kinetic studies suitable patterns are obtained in about 20 sec, so that subsequent exposures can be made in intervals of 40 - 60 s.

c) Statistical accuracy of PSD data, reduction of background

Because of the larqe number of pulses the statistical accuracy of PSD data is better than that of conventional data which means that a smoother background is obtained with PSD measurements. Sometimes the unwanted incoherent background of PSD data can be reduced by placinq thin Al-foils in front of the counter. Apart from using Al-foils in front of the PSD, the proper choice of wave lenqth (e.g. Co for Fe-containinq samples) is advisable.

In cases of fibres, sinqle crystals, micro samples and capillaries the proper choice of collimators and brass masks helps considerably to reduce unwanted incoherent radiation and air scattering. In bad cases He-tunnels might improve the background situation.

Fiq. 3 and 4 show PSD diagrams of mine samples and polymers. In Fig. 3 two PSD scans of mine samples (1,0 mg dust on filters, 3 min exposed) are shown which were used for quantitative analyses and in Fig. 4 the recrystallization of low molecular polyethylene (molecular weight about 3000) from melt is shown.

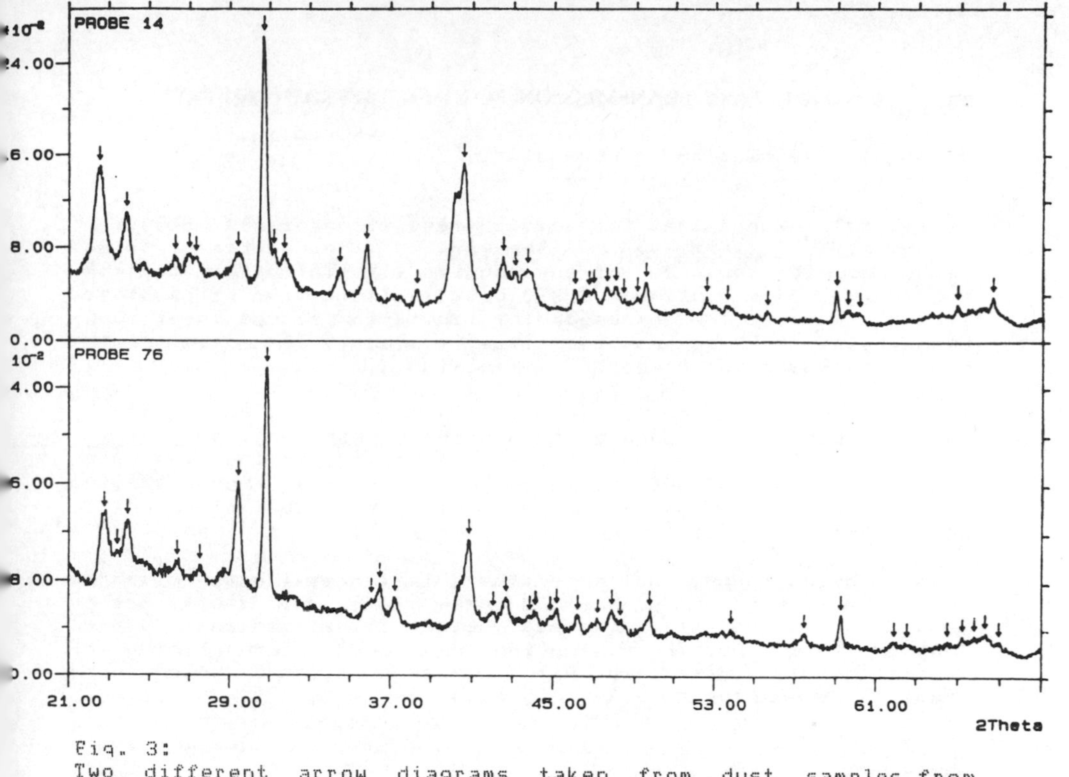

Fig. 3:
Two different arrow diagrams taken from dust samples from coal mines (180 s exposed, about 1 mg of material on filter)

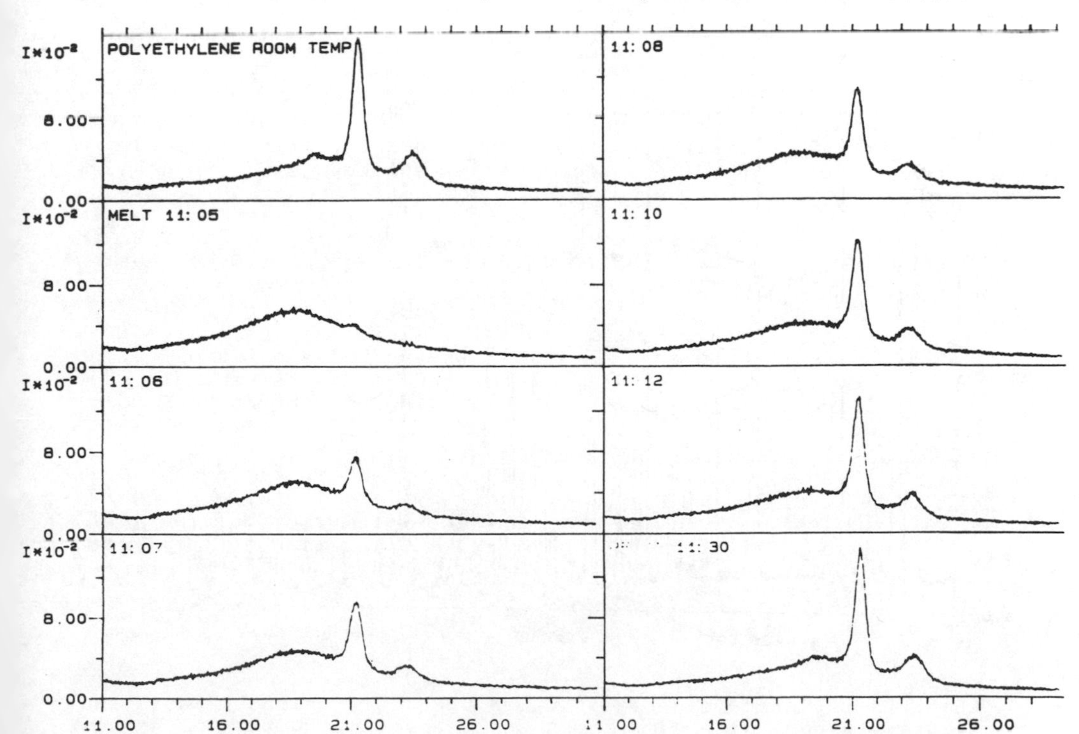

Fig. 4: The recrystallization process of low molecular polyethylene from melt (30 s exposed, temperature gradient about 20 degrees/ min).

d) Angular resolution of PSD-data

It is well known that the angular resolution of powder pattern is primarily determined by the width of the counter slit. Because of the absence of any counter slit infront of the PSD, the angular resolution of PSD data is limited to Delta 2theta = 0.15 degree. The corresponding figure for conventional counters using 0.2 mm counter slits is about 0.10 degree for the beam path which was described in section I.

In this context it should be mentioned that angular resolution means that neighbouring peaks can be identified by a quick peak search routine and indicated with arrows on the arrow diagrams (see fig. 3 + 5). It is evident that a doublet can be already visually identified at a lower angular resolution.

It has been found, that a considerable increase of angular resolution over ranges of +/- 5 degrees of PSD data can be reached by increasing the distance between specimen and PSD from 130 mm to 200 mm. The reason for this is the increase of electronic angular resolution with increasing distance which overcompensates defocusing beam path effects.

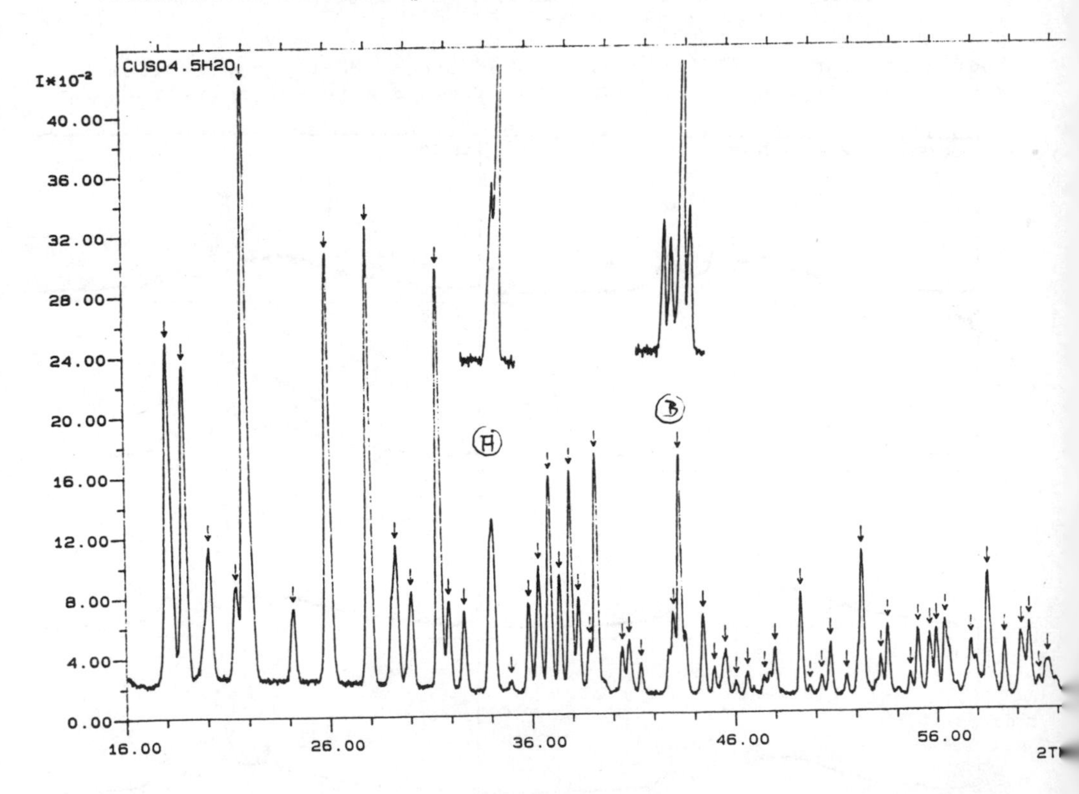

Fig. 5: PSD arrow diagram of $CuSO_4.5H_2O$ (Co K alpha 1 radiation, 180 s exposed). The sections A and B are shown with high angular resolution in the upper part of Fig. 5 (distance sample to PSD: 200 mm, 180 s exposed)

An example is qiven in Fiq. 5 where the PSD diaqram of triclinic $CuSO_4.5H_2O$ is shown taken with Co K alpha 1 radiation in 3 minutes. There are two angular regions with insufficient angular resolution. A) (a broader line which should be at least a doublet) and B) (which is a unresolved quartet). The well resolved sections A and B are shown in the upper part of Fig. 5 whereby the distance between the PSD and the speminen was 200 mm (3 min exposed). The new peak position of A and B can be routinely transferred to the original diagram where the old peak positions are deleted. Correct indexing of $CuSO_4.5H_2O$ was possible with these improvements. From many applications it can be concluded that the described procedure has to be made only in few cases of heavily overlapped regions.

e) Lifetime of PSD, reproducibility and linearity of PSD data

The Stoe made PSD is filled with a Ar + 10 % CH_4 mixture, the normal pressure being 5 atm. In normal use there is no need for refilling. Sometimes, however, pressure decreases over weeks from 5 atm to 4 atm are observed and because of the lower quantum efficiency a refilling of the PSD is advisable which is easily carried out in a few minutes. Apart from that the PSD needs no service and its lifetime is unlimited. The maximum counting rate is about 10 **5 pulses per second and pulse rates from powder samples with the beam path of Fig. 1 are far beyond this limit.

Careful studies have confirmed a linearity of PSD data in intensity of 0.1%. It has also been confirmed that the electronic scale which is coincident with the 2theta scale is constant over long periods of time because the electronic components of the PSD counting chain and of the PSD interface are extremely well thermostabilized.

f) Sample preparation

It was found that optimal sample preparation is extremly important for obtaining proper PSD data. Transmission samples should have the right thickness, the material should be of homogeneous particle size and there should be no preferred orientation.

In order to satisfy these conditions, only sieved fractions should be used for precision measurements. The TAC display allows to observe easily strong reflections from larger crystallites.

For transmission samples precisely made circular rotatable sample carriers are provided and the height of the outgoing radiation is limited by a circular brass mask which can be inserted and fixed to the carrier. The amount of material should be chosen in such a way that the transmission factor Io/I should be in the range between 4 and 8. Low absorbing material (10 - 30 mg) is suspended between two thin polyacetate foils, the brass mask of suitable diameter is inserted and before tightening the mask, the material is handpressed against the mask thus forming a layer of uniform thickness.

High absorbing brittle material cannot be prepared in this way because it would not held its position during rotation. In such cases a small amount (0.1 - 3 mg) is fixed on a polyacetate foil with a fraction of a droplet of a water soluble glue or with vaseline. A spanning device is available which makes the preparation of thin layer samples easier.

It is also important to control the mu . r-factor of capillaries. If even thin capillaries show high absorption factors, thick capillaries should be used which are mounted in such a unsymmetrical way that the rectangular X-ray beam just strikes the capillary tangentially in a quasi reflecting position.

IV. Fields of application

From the above sections four main advantages of the novel diffractometry became evident:

- 'Life' observable diffraction pattern.
- Exposure time between 20 - 180 s.
- Highly resolved precise and reproducible PSD data.
- Favourable statistical accuracy of PSD data.

This allows a wide range of applications in various fields of powder diffractometry. Our own experiences have been concentrated to the following fields:

- Search matches of multiphase mixtures with JCPDS and own standard files.
- Indexing of pure phases.
- Quantitative analyses of airborn samples (see JAC (1981) 14, 291 - 296).
- Kinetic studies of solid state reactions and phase transitions.
- Polymer studies.
- Analyses of micro samples (about 0.1 mg).

Apart from these conventional applications the new method opens a 'POWDER DIFFRACTOMETRY WITHOUT A DIFFRACTOMETER' since the PSD can be operated stationary and because it can be calibrated to the true 2theta scale without a diffractometer. Thus studies of chemical reactions in chemical cells of various designs are possible. Because of the transmission geometry there is no need for a planar specimen. For the same reasons the new method can be used in the future for controlling various chemical processes.

Because of the limited space of this survey the software which is available was not discussed but it certainly became evident that for the different fields of applications a most advanced software package has to be provided to take advantage of the novel diffractometry. This package ranging from peak search, pattern decomposition and indexing to search match and parameter refinement in conjunction with a most advanced grafic system has been developed at Stoe & Cie GmbH under the direction of H. Langhof by a team of experienced scientists and will be continuously further developed. Descriptions of the current state of the software package are available on request.

COMPUTER ANALYSIS OF POWDER DIFFRACTION DATA AND STRUCTURE REFINEMENT USING RIETVELD METHODS

BY

ROBERT A. SPARKS
NICOLET INSTRUMENT CORPORATION

INTRODUCTION

In the last decade the cost of the minicomputer has decreased to the point that it has become economical to attach such computers to powder diffractometers in order to control the collection and analysis of data. As a result, all of the manufacturers of powder diffractometers now offer automation packages consisting of computer hardware and program libraries. Most of the existing automated diffractometers allow only one job to execute at a time (i.e. either data collection or one data analysis program). As the cost of the minicomputer (and especially the cost of memory) decreases still further, timesharing of jobs becomes economical.

In many of the analysis procedures discussed below the ability to see a powder diffraction pattern and to interact with that pattern is very convenient. Therefore, many automated systems have interactive graphics display hardware.

To be able to store programs and data sets, floppy diskette drives are used. The Joint Committee for Powder Diffraction Studies (JCPDS) powder diffraction file when condensed occupies over 7 Mbytes of storage. A magnetic tape drive or preferably a magnetic disk drive must be present to be able to search this data set.

A typical hardware configuration is shown in figure 1.

Most of the programs are written in BASIC or FORTRAN. Some of the manufacturers provide source programs, interpreters, compilers and linkers so that the users can modify the algorithms if they wish.

DATA COLLECTION

Data is usually collected in a step scan mode. The theta and two-theta axes are positioned with stepping motors (or one motor if the axes are coupled); the diffracted x-ray intensity is measured for a given interval of time, and the axes are then stepped to the next sample position. The Nicolet data collection program is interactive. Two-letter commands let the user specify variables such as two-theta ranges, step sizes, count times, file names for storage of data, plot parameters for a strip chart recorder, and which samples in a sample changer are to be used. A separate file is created for raw data from each sample. Subsequent processing programs can then be used on each of these files. Doyle (1979) modified this program so that it will do a fast scan of the data. Based on the peak positions and intensities it finds it then scans only over those regions where reflections are present. Count times for each region are set so that about the same number of counts will be measured for each reflection.

PEAK FINDING

Many of the analysis procedures require first finding accurate values of d-spacings and intensities. The most widely used technique is that of Savitsky and Golay (1964). The method assumes that the data has been collected at equal increments in two-theta. For each data point, a least squares low order polynomial is fit to it and its neighboring data points. For example, a common algorithm fits a cubic to the five data points with the given data point in the center. The second derivatve of the polynomial is evaluated at the given data point. The procedure is repeated for every data point (the data must be extended at the ends of the two-theta range). Peaks are assumed to occur where the second derivatives are most negative. An interpolated position of the peak maximum is calculated by setting the first derivative of the polynomial to zero. The biggest problem is in setting a bound on the value of the second derivative to determine what will be considered to be a reflection and what will be considered to be noise in the data. It has been observed that for a given choice of value for the bound, there will be a tendency for real peaks to be missed at high two-theta values and fake peaks chosen at low two-theta values (Byram and Christensen (1981)). Therefore, the bound is multiplied by a weighting function which is determined empirically.

The algorithm is sensitive to noisy data and sometimes it is advantageous to smooth the data first. The methods of Savitsky and Golay (1964) can also be used for this purpose.

It is also useful to remove background as well as possible before using a peak locating algorithm. Mallory and Snyder (1980) have devised a method which determines which data points to assign to background and then calculates a spline fitting these points. The spline passes smoothly under the reflection peaks and is subtracted to produce a background-free data set.

The second derivative method works well in regions where there is not too much overlap of peaks. It finds peak positions very accurately where there is no overlap. For overlapped peaks it tends to produce peak positions which are too close together and intensities which are too large. The algorithm finds both $K_{\alpha 1}$ and $K_{\alpha 2}$ peaks. An option of the version written by Byram and Christensen (1981) will subtract the $K_{\alpha 2}$ peak and add its intensity to the $K_{\alpha 1}$ peak.

From the wavelength of the incident x-radiation the d-spacings are calculated and stored together with peak intensities in a file which can be used for further processing.

PATTERN DECOMPOSITION

The pattern decomposition method is better than the second derivative method for determining d-spacings and intensities in regions where there is considerable overlap of peaks. If the peak shape as a function of two-theta is known, then it is possible to fit the entire powder pattern with the appropriate number of peaks. There are three variables which are determined for each reflection (two-theta, integrated intensity, and peak-width). It is also necessary either to subtract a known background function or include background variables in the model. The power of the pattern decomposition method lies in the fact that all of the data points are used instead of only those near peak maxima as in the second derivative method.

Unfortunately, the peak shape in x-ray powder diffraction is not a simple analytical expression. The shape is a function of of the x-ray K_{α} spectral line profile passed by the monochromator, the instrumental abberations of the diffractometer, and the broadening effects of the powder specimen. Taupin (1973) and Huang and Parrish (1977) use three Lorentzians to fit the $K_{\alpha 1}$ peak, three to fit the $K_{\alpha 2}$ peak and a seventh Lorentzian to fit the weak $K_{\alpha 3}$ satellite group. For each Lorentzian three parameters (two-theta, integrated intensity and peak width) are adjusted to give as good a fit as possible with the observed data.

Schreiner and Jenkins (1982) used a composite Lorentzian plus a small symmetric Lorentzian at the low two-theta end of the peak to fit the $K_{\alpha 1}$ peak and an independent composite Lorentzian plus a symmetric Lorentzian to fit the $K_{\alpha 2}$ peak. A total of twelve parameters are determined in this method.

Howard and Snyder (1982) have evaluated several mathematical models for peak shape (Table 1). The Nicolet method uses as the peak shape a table of smoothed (see above) intensity values measured for long count times at increments of 0.01 degree in two-theta. Linear interpolation is used between table entries.

Regardless of the model for the peak shape it is necessary to measure isolated (no reflections overlapping) peaks from standard specimens which are well-crystallized (low mosaicity) randomly oriented samples. A series of reflections are chosen so that the whole two-theta range is adequately covered. Examples of standards are quartz, silicon, tungsten, and Linde A zeolite (for low two-theta reflections). Linear interpolation of the peak shape parameters or tables is used for two-theta values between those actually measured.

Once the calibration procedure has determined model parameters or tables, these are used for all subsequent analysis of unknown samples. It is important that if any physical changes in the instrument are made, the calibration procedure must be redone.

The program written by Sparks (1979) for pattern decomposition (PROF) divides the pattern into two-theta regions that are bounded on each end by background. The program tries to minimize the following function in each region:

$$\chi^2 = \sum \left[\frac{I_{oi} - I_{ci}}{\sigma_i}\right]^2 \Big/ (N-n)$$

where I_{oi} is the measured intensity value, I_{ci} is the corresponding calculated value, σ_i is the corresponding experimental standard deviation (Poisson statistics are assumed), N is the number of data points and n is the number of parameters which are allowed to vary. For each region, PROF calculates a linear background function which is then subtracted from the observed data. From the calibration file a peak is fit to the observed data. Two-theta, the integrated intensity, and the full-width-half-maximum (FWHM) are adjusted until χ^2 is minimized. This peak is then subtracted from the observed data. A second and subsequent peaks are added until no remaining data point is larger than some preset value (default is four standard deviations). Finally a non-linear least squares to minimize χ^2 is performed. The variables are the two background values, one at each end of the region, and for each peak the three values--two-theta for $K_{\alpha 1}$, integrated intensity, and FWHM. The program can handle up to 15 independent peaks in each region. The program prints χ^2 and all of the refined variables together with their estimated standard deviations.

In a test case of a mixture of barium chloride dihydrate, potassium iodide, silicon, and zinc oxide, data was collected at steps of 0.04 degree in two-theta. The maximum count was 1993. PROF found a total of 70 reflections. In the region from 34.04 to 38.72 degrees, PROF found ten overlapped peaks. (See figure 2.) The strongest and weakest reflections have integrated intensities of 6486 and 454, respectively. The estimated standard deviations in two-theta are 0.0013 degree and 0.009 degree, respectively. The estimated standard deviations in integrated intensity are 1.6 percent and 7.7 percent, respectively. Much better precision can be obtained with longer count times, but systematic errors will tend to dominate (Huang and Parrish (1979)).

In cases of severe overlap there will be more than one way to decompose a given region. Thus, for the 53.0 to 58.0 degree region of kaolinite and silicon (added as an internal standard) it was found that four peaks with FWHM of 0.17 degree, 0.38 degree, 0.96 degree and 1.37 degree fit as well as one peak with FWHM of

0.175 degree and 12 peaks with FWHM of 0.30 degree. (See figure 3.) Clearly, additional information is needed to decide which alternative to choose. To help guide PROF, an interactive capability was added to the program. The user can specify the number of peaks and their approximate locations, intensities, and FWHM's. PROF will then refine these parameters.

INTERNAL STANDARDS

Often it is difficult to mount a sample so that it will be centered in the diffractometer. By adding an internal standard with reflections at known two-theta positions, it is possible to correct all of the unknown two-theta values which have been determined by the second derivative or pattern decomposition method. INSTD is a program which fits a low order polynomial through the known two-theta positions and then corrects all other two-theta positions. It has been found that a first or second order polynomial is sufficient for this purpose.

SEARCH/MATCH

Once d-spacings and intensities have been determined, several different processing programs can be used. The most popular are the various search/match programs which are used to identify the phases present in the unknown sample. The programs of Johnson and Vand (1967), Nichols (1966) and Frevel (1965) were written to be run on large computers. Newer programs have been written for minicomputers. To make these programs efficient, the large standards files of the Joint Committee for Powder Diffraction Studies (JCPDS) have been condensed. Thus, as condensed by Sparks (1982), the Inorganic, Organic, and Mineral files require 5.1, 2.1 and 0.7 Mbytes, respectively. Many of the programs search the complete standards file or selected sub-file evaluating a figure of merit for each standard. At the end of the search, standards with the highest figure of merit are printed. The SANDMAN program (North American Philips) (Schreiner and Surdukowski (1982)) picks out those patterns from this list which in combination give the best account of the lines in the unknown.

The program, SEARCH, written by Sparks (1982), uses a strong line directory which is ordered by d-spacing of the most intense line in each standard pattern. Also included in this directory file are the d-spacings of the second and third most intense lines. The chemical elements and chemical groups present in the standard are also included in this directory file. For the n most intense lines of the unknown (n is typically set to 10) the directory file is searched. A window in two-theta is set (typically ± 0.1 degree). For each of the n lines all those entries in the directory that match within the window are tested. The second and third most intense lines of the standard must also be present in the unknown. The chemistry of the standard must also match that specified by the user. If all of these criteria are met, the complete standard pattern is retrieved from the main standard file and a figure of merit calculated. Optionally the program will subtract off the "best" standard and repeat the process until no further standards match the remaining lines. The advantage of using this strong line directory is that the search algorithm goes very fast because only a small fraction of the patterns in the complete file are examined. Typically a search of the Inorganic file takes about 10 minutes. The figure of merit is a linear function of the following parameters: (1) the fraction of the total intensity of the sample accounted for by the total scaled intensity of the matched lines of the standard, (2) the fraction of the total intensity of the standard accounted for by the matched lines of the standard, (3) a correlation coefficient of the intensities of the sample and the standard for the matched lines, (4) the average two-theta error for the matched lines of the standard. The user specifies the four coefficients of the linear function. The default coefficients are 1, 0, 0, 0. For specimens with large amounts of preferred orientation it has been found necessary to give large weight to the fourth parameter. Best results are obtained when chemical information is used and when accurate d-spacings have been measured.

QUANTITATIVE ANALYSIS

From accurate intensity measurements it is possible to determine the percent composition of the phases in a mixture. In some procedures it is necessary to know the mass absorption coefficients of the phases present. The algorithms are simple (Klug and Alexander (1974)). Usually many samples with some standards are placed in a sample changer and all of the samples are measured and analyzed automatically. The procedures and programs tend to be tailored for each individual type of analysis. Hubbard, Robbins and Snyder (1982) describe the set of programs used at the National Bureau of Standards.

INDEXING

From accurate d-spacings for a single phase powder sample, it is possible to assign indices to the diffraction lines and calculate unit cell dimensions. All of the various indexing programs are based on trying to solve the following set of equations for the observed d-values:

$$1/d^2_{hk\ell} = Q_{hk\ell} = h^2Q_A + k^2Q_B + \ell^2Q_C + k\ell Q_D + \ell h Q_E + hk Q_F$$

where Q_A, Q_B, Q_C, Q_D, Q_E, and Q_F are functions of the reciprocal lattice parameters and h, k, and l are integer indices. Shirley (1979) has shown that only the first 20-30 lines are needed but that these need to be measured accurately. Low symmetry (triclinic and monoclinic) problems can be solved but take more computer time than the high symmetry problems. A review of the various indexing programs is given by Shirley (1978). A powerful identification procedure is: (1) determination of lattice parameters, (2) transformation to a reduced cell, and (3) search of the Crystal Data file.

SOLUTION OF CRYSTAL STRUCTURES

Small crystal structures can be solved and refined using only powder diffraction data. Using the pattern decomposition program PROF described above, Calvert, et al. (1984) showed that lattice parameters could be obtained with a precision of 1:30000; two-theta angles with a precision of 0.006 degree to 0.013 degree, and intensities with a precision of 2 to 6 percent. They solved the structure of $CaNi_5H$ by using PROF to obtain intensities for 30 resolved reflections. After a cycle of least squares refinement, they extended the data set to include reflections which were overlapped in the powder pattern. For the overlapped reflections, intensities were assigned based on ratios of calculated intensities. Refinement of 11 position parameters, 6 isotropic temperature parameters and one scale factor led to an R value of 0.069.

Clearfield (1984), also using PROF to obtain integrated intensities, solved a nine-atom inorganic stucture by the heavy atom Patterson method. He refined the structure with the Rietveld program by Baerlocher and Hepp (1979).

Will (1979) has written a powder least-squares program, POWLS, which minimizes the function

$$\sum_i w_i (G_{oi}^2 - G_{ci}^2)^2$$

with respect to atomic position and temperature parameters where w_i is the weight for the ith unresolved group of reflections

$$G_{oi}^2 = \sum_j I_{oij} / L_{ij} P_{ij} \quad ,$$

I_{oij} is the intensity of the jth relection in the ith group, L_{ij} is the corresponding Lorentz correction, P_{ij} is the corresponding polarization correction, and the summation is over all reflections in the ith unresolved group in the powder pattern. The observables are thus the corrected integrated intensities of those regions which are not overlapped. G_{ci}^2 is the ith corresponding calculated value which is a function of the atomic parameters in the structure.

RIETVELD ANALYSIS

Rietveld (1967) wrote a least squares program to refine crystal structure parameters using neutron powder diffraction data. In the last few years this method has been extended to use x-ray diffraction data. The quantity minimized is:

$$\sum_i w_i \left| y_{oi} - y_{ci} \right|^2$$

where y_{oi} is the intensity observed at the ith step in the step scan. The y_{ci} is defined as

$$y_{ci} = s \sum_k p_k L_k P_k \left| F_k^2 \right| G(x_{ik}) Q_k + y_{bci}$$

where s is the scale factor, L_k and P_k are the Lorentz and polarization factors for the kth Bragg reflection, F_k is the structure factor, p_k is the multiplicity factor, Q_k is the preferred orientation function, x_{ik} is the Bragg angle for the kth reflection, $G(x_{ik})$ is the reflection profile function and y_{bci} is the background function. The parameters that are adjusted by the non-linear least squares procedure are atomic positions, atomic thermal parameters, atomic site occupancies, scale factor, overall thermal parameter, lattice parameters, 2θ - zero correction, reflection profile parameters, preferred orientation parameters, and background function parameters. As pointed out above, x-ray reflection profiles cannot be expressed as simple analytical functions. Many different functions have been used (Young (1979)). The non-linear least squares algorithm is the same as used in single crystal least squares with the following exceptions: (1) The Rietveld program must include code for the additional types of parameters. (2) The individual reflection intensities are spread over many data points (dependent on the reflection profile function). As a result the Rietveld program has more program statements than the single crystal least squares program. The total number of parameters refined is less for a Rietveld program (less than 100) because of severe overlap of reflections and hence lack of information at high two-theta values. Anisotropic thermal parameters could be refined for very small structures, but for larger structures there is not enough information in the data to produce meaningful vibrational parameters. One of the most widely used programs was written by Wiles and Young (1981).

Lattice parameter cell constants from Rietveld analysis usually have high precision. The Rietveld method gives reasonable atomic position parameters, but thermal parameters are less satisfactory. The standard deviations obtained from the Rietveld method have been generally regarded as too small. Prince (1984) has shown that these standard deviations are measures of precision--not necessarily accuracy. It is apparent that the models need to be improved to include all of the experimental parameters.

OTHER POWDER DIFFRACTION PROGRAMS

From line broadening it is possible to measure crystallite size and lattice strain (Langford (1979)). Changes of d-spacings for certain reflections can give information about applied or residual stresses (Cohen, Dölle and James (1979)). Changes in lattice parameters and hence d-spacings for some reflections can be used to analyze solid solutions. Special programs for these applications have been written (Anderson (1984)).

CONCLUSION

Recent developments in powder diffraction include widespread use of the minicomputer for both collection and analysis of data, more powerful algorithms, and solution and refinement of crystal structures from powder data.

REFERENCES

Anderson, P., (1984). Private communication, Kaiser Aluminum, Pleasanton, California, USA.

Baerlocher, Ch and Hepp, A. (1979). Accuracy in Powder Diffraction, National Bureau of Standards Special Publication 567, US Government Printing Office. 165.

Byram, S. and Christensen, A. (1981). Operating Instructions for PEAK. L11 User's Manual, Nicolet Instrument Corporation. Madison, Wisconsin, USA.

Cohen, J.B. Dölle, H. and James, M.R. (1979). Accuracy in Powder Diffraction, National Bureau of Standards Special Publication 567, US Government Printing Office. 453-477.

Calvert, L.D., Murray, J.J., Gainsford, G.J., and Taylor, J.B. (1984). Mat. Res. Bull. 19, 107-113.

Clearfield, A. (1984). To be published. Journal of Inorganic Chemistry.

Doyle, J.H. (1979). Materials Technology R & D Report PMRB-79-031 Rockwell International Energy System Group, Golden, Colorado, USA.

Frevel, L.K. (1965). Analytical Chemistry 37, 471-482.

Howard, S.A. and Snyder, R.L. (1982). Advances in X-Ray Analysis vol. 26 (Plenum Press, New York, 1982), 73-80.

Huang, T.C. and Parrish, W. (1977). Advances in X-Ray Analysis, vol. 21 (Plenum Press, New York, 1978), 275-288.

Huang, T.C. and Parrish, W. (1979). Accuracy in Powder Diffraction, National Bureau of Standards Special Publication 567, US Government Printing Office, 95-110.

Hubbard, C.R., Robbins, C.R. and Snyder, R.L. (1982). Advances in X-Ray Analysis, vol. 26 (Plenum Press, New York, 1982) 149-156.

Johnson, G.G. and Vand, V. (1967). Ind. Eng. Chem. 59, 19-31.

Klug, H.P. and Alexander, L.E. (1974). X-Ray Diffraction Procedures for Polycrystalline and Amorphous Materials, second edition (Wiley and Sons, New York, 1974).

Langford, J.I. (1979). Accuracy in Powder Diffraction, National Bureau of Standards Special Publication 567, US Government Printing Office. 255-269.

Mallory, C. and Snyder, R.L. (1979). Accuracy in Powder Diffraction, National Bureau of Standards Special Publication 567, US Government Printing Office. 93.

Nichols, M.C. (1966). A FORTRAN II Program for the Identification of X-Ray Powder Diffraction Patterns, UCRL-70078, Lawrence Livermore Laboratory. Livermore, California, USA.

Prince, E. (1984). "Precision and Accuracy in Structure Refinement by the Rietveld Method," to be presented at the IUCR meeting in Hamburg.

Rietveld, H.M. (1967). Acta Cryst. 22, 151.

Savitsky, A. and Golay, M.J. (1964). Analytical Chemistry 37, 1627-1639.

Schreiner, W.N. and Jenkins, R. (1982). Advances in X-Ray Analysis, vol 26 (Plenum Press, New York, 1982), 141-147.

Schreiner, W.N. and Surdukowski, C. (1982). J. Appl. Cryst. 15, 513-523.

Shirley, R. (1978). Computing in Crystallography. eds H. Schenk, R. Olthof-Hazenkamp, H. van Koningsveld and G.C. Bassi. Delft University Press, Delft) 221-234.

Shirley, R. (1979). Accuracy in Powder Diffraction, National Bureau of Standards Special Publication 567, US Government Printing Office, 361-382.

Sparks, R.A. (1979). Iterative Profile Analysis. L11 User's Manual. Nicolet Instrument Corporation. Madison, Wisconsin, USA.

Sparks, R.A. (1982). Nicolet XRD Search/Match Programs. L11 User's Manual. Nicolet Instrument Corporation. Madison, Wisconsin, USA.

Taupin, D. (1973). J. Appl. Cryst. 6, 266-273.

Wiles, D.B. and Young, R.A. (1981). J. Appl Cryst. 14, 149-151.

Will, G. (1979). J. Appl. Cryst. 12, 483-485.

Young, R.A. (1979). Accuracy in Powder Diffraction, National Bureau of Standards Special Publication 567, US Government Printing Office. 143-163.

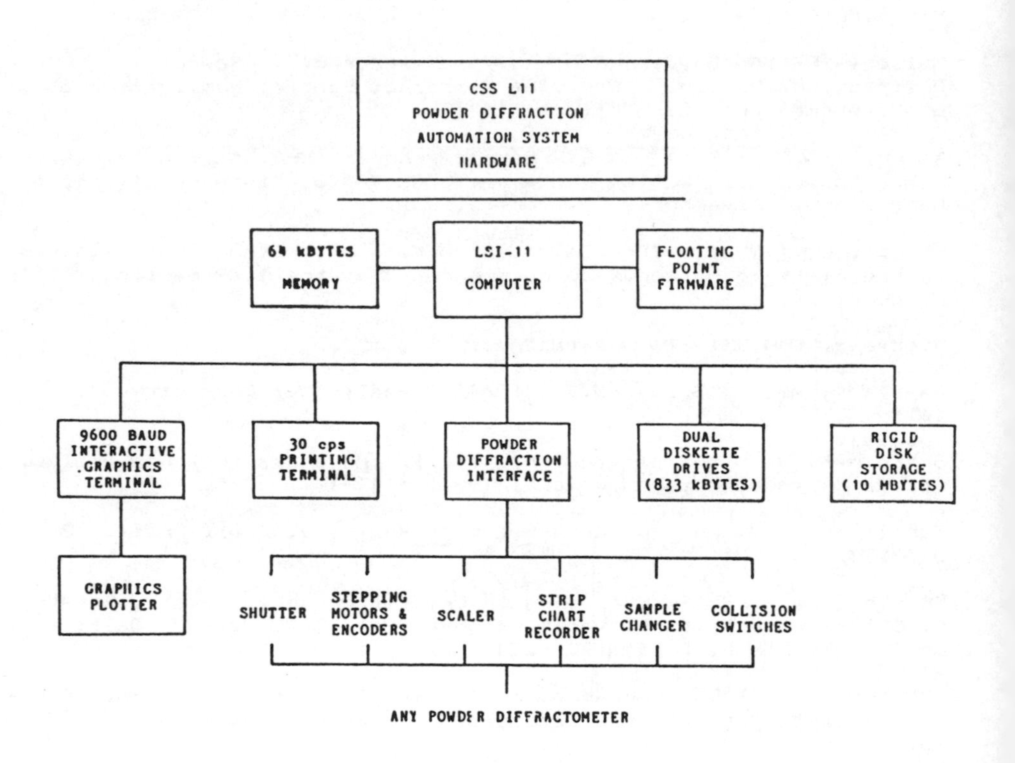

Figure 1

Table 1. Profile Functions

#	Name	Form: I(x) =	Conditions:
1	Lorentzian	$I_0/(1+kx^2)$	$k = 1\ /\ (FWHM/2)^2$
2	Mod. Lorentzian	$I_0/(1+kx^2)$	$k = 0.4142/(FWHM/2)^2$
3	Int. Lorentzian	$I_0/(1+kx^2)^{1.5}$	$k = 0.5874/(FWHM/2)^2$
4	Pearson VII	$I_0/(1+kx^2)$	$k = (2^{(1/m)}-1)/(FWHM/2)^2$
5	Split Pearson VII	$I_0/(1+kx^2)^m$ $I_0/(1+k'x^2)^{m'}$ where	for x positive for x < or = 0 $k = (2^{(1/m)}-1)/(FWHM/2)^2$ $k' = (2^{(1/m')}-1)/(FWHM/2)^2$
6	Gaussian	$I_0\exp(-kx^2)$	$k = 0.6931/(FWHM/2)^2$
7	Voigt	Lorentzian*Gaussian	Convolution product

Where:		
	$x = 2\theta_i - 2\theta_k$	Distance from the Bragg Angle($2\theta_k$)
	FWHM	Full width at half maximum intensity
	m, m'	Shape factors

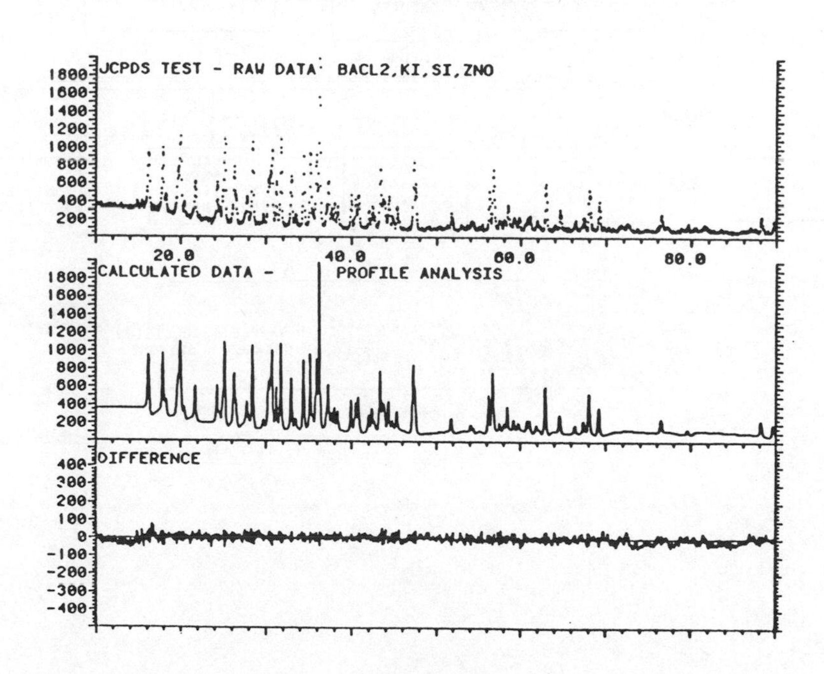

Figure 2

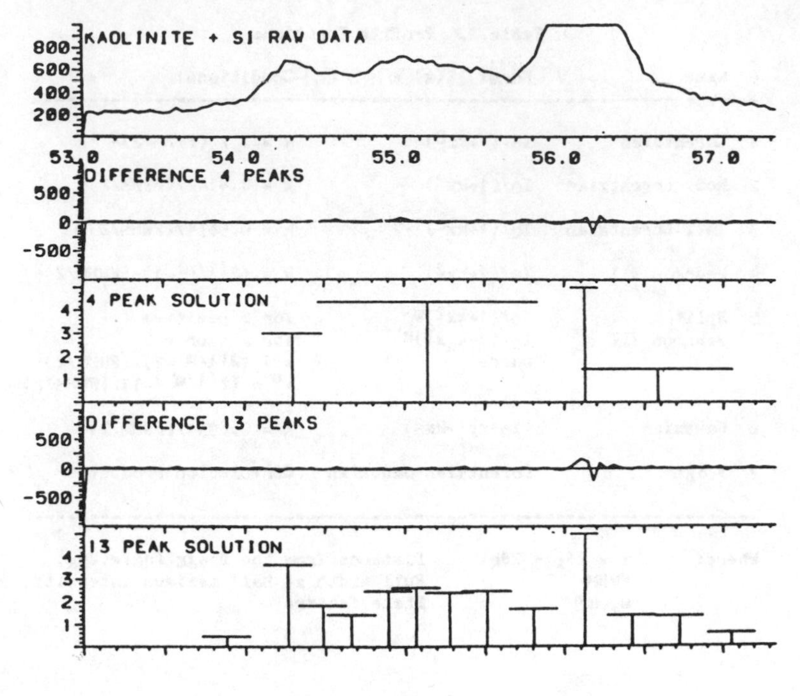
KAOLINITE + SI RAW DATA
800
600
400
200
53.0
54.0
55.0
56.0
57.0
DIFFERENCE 4 PEAKS
500
0
-500
4 PEAK SOLUTION
4
3
2
1
DIFFERENCE 13 PEAKS
500
0
-500
13 PEAK SOLUTION
4
3
2
1

Figure 3

Crystallographic databases

THE INORGANIC CRYSTAL STRUCTURE DATA BASE

G.Bergerhoff

Institute for Inorganic Chemistry, University of Bonn
Gerhard-Domagk-Str. 1, D - 5300 Bonn 1

Any scientific work should not and could not be done without knowledge of the results of our predecessors. But the enormous amount of results induces that we neglect the results instead of using them. The computer which helped us in the last years to gain these results, should now also help us to use them in the best way. This is especially true in a field like crystal structure analysis because here scientists evaluated a lot of complex constructed and now widely spread data. Olga Kennard and her coworkers demonstrated with the Cambridge Crystallographic Data File what the computer can render in this field.

I want to stress that such work should be expanded to many other fields of science because the setup of data bases not only enables us to make use of the data in the best way but also obliges to consider our data critically in many respects. To fulfil these tasks, data bases should be 1. complete, 2. up-to-date, 3. correct, 4. versatile and 5. user-friendly.

The Cambridge Crystallographic Data File omits the pure inorganic compounds for several reasons and thus we had the opportunity to demonstrate how we try to fulfill the cited five conditions for a good data base (Bergerhoff et al. (1983)).

To be complete:
Once the user has recognized the advantages of a numerical data base he wants to have all data in one place. Thus we try to be complete for inorganic structures in respect to the following principles. The ICSD contains all

1. Structures which have no C-C-, no C-H-bonds in any one of their residues. (Cambridge has the opposite definition). Structures which have incorporated at least one of the

non-metallic elements. This means: H, He, B, C, N, O, F, Ne, Si, P, S, Cl, Ar, As, Se, Br, Kr, Te, I, Xe, At, Rn. (the metals file (Calvert (1981)) has in principle the opposite definition)

2. Structures the atomic coordinates of which have been fully determined. Coordinates of hydrogen and vagabond ions like in zeolites can be missed.
3. Structures to which refer Strukturbericht, Structure Reports, Crystal Data and Landolt-Börnstein.

This aim has been fully reached in respect to the bibliographic part of the 31.180 datasets. 19.552 of them also contain the full numerical information. (The numbers for the organic file are 42.377 resp. 34.742.) In spite of these numbers I am shure we have not all known structures. Please check the data base and help us to make it complete.

To be up-to-date:

Data bases in principle have the possibility to be up-to-date if the new results reach them without delay. Many references we get through regularly scanning of Chemical Abstracts and of the ten most important journals

Acta Crystallographica, American Mineralogist, Doklady Akademii Nauk USSR, Journal of Solid State Chemistry, Kristallografiya, Materials Research Bulletin, Zeitschrift für anorganische und allgemeine Chemie, Zeitschrift für Kristallographie, Zeitschrift für Naturforschung, Zhurnal Strukturnoi Chimii.

But the input could be much more up-to-date and complete if it would be possible to find an arrangement how data could reach directly the data base producers without the long way through typescript - published paper - reference journal. It is a good thing that several journals such as Acta Cryst., J.chem.Soc. Chem.Comm., Angewandte Chemie already cooperate with the data base producers to provide them with data which normally will not be published. The annual increase of the ICSD is about 1200 structures but the time lag may be one year and more.

To be correct:
Information which is not correct is worthless. This is especially true for data bases. The user not only wants to work with correct data but also a successful retrieval process depends on correct data. There are three kinds of errors.
1. Input errors, which are our fault and should be avoided.
2. Printing errors, which are the author's faults and are eliminated by correspondence.
3. Errors by misunderstanding of the authors intention. They make most trouble and could be avoided by strictly using the recommended standard data structure.

In any case it is impossible to eliminate all errors by means of inspection by eye. Therefore it is highly necessary to develop programs which allow the computer to check as many data as possible. For non-numerical data these programs are in development. Numerical data are strictly checked for consistency where the program uses the following scheme.

```
                        Structured formula  -----!
                                                 !---------->
                        Number of formula units -!          !
                                                            !
                     !--Atomic symbol -----------!          !
                     !                           !          !
                     !--Number of positions -----!---> Molecular
Electroneutrality <--                            !       formula
                     !--Site occupation ---------!          !
                     !                                      !
                     !--Oxidation state                     !
                                                            !
                        Density------------------<          !
                                                 !----------<
                     !--Unit cell dimensions ---->
                     !
Atomic   <-----------!--Atomic coordinates
 distances           !
                     !--Space group
```

The check results are included into the data base. Typical examples are: calculated density anomalous, given molecular formula different from formula calculated from number of positions and site occupation for each atom, no electroneutrality, temperature factors not plausible, atomic distances largely deviate from sum of ionic radii. Once more users of the data base are asked to help to correct data when they are able to do so.

To be versatile:
Once all data have been collected and corrected as far as possible the data base should answer all imaginable questions of the user. And this should be done in the most convenient manner: in a dialog retrieval at the computer terminal. For this reason Rolf Sievers and Rolf Hundt at our institute developed the retrieval system CRYSTIN. The system not only allows the access to the ICSD but also handles simultaneously the Cambridge and the metals file in such a manner that the user does feel all crystallographic data bases as one unit. Thus the border lines between the crystallographic data bases, cited above, no longer exist for the user of CRYSTIN and he must not decide which area of chemistry his compound belongs to.

To formulate queries to the data base the user can combine descriptors of various types by logical connections. The following table of descriptors gives the possibilities existing today.

Chemical data
- Chemical element
- Chemical element and oxidation state
- Chemical element and stoichiometric index of molecular formula
- Element group
- Part of name and part of formula (by string search)
- Number of different elements
- Mineral name
- Origin of minerals (by string search)

Refcode (Cambridge Crystallographic Data File only)
Chemical Class (Cambridge Crystallographic Data File only)

Structural data
Unit cell volume
Formula type (e.g. AB, A2B3, A2BX4)
Pearson symbol
Defect structure (in general or polytype, twinned and modulated structures, mixed crystals)
Shortest distance between two atoms in a structure

Symmetry data
Space group (by Hermann-Mauguin symbol or number of IT)
Crystal class
Laue class
Crystal system
Polarity and centrosymmetry

Methodic data
Reliability index
Diffraction method
Temperature and pressure of measurement
Test result

Bibliographic data
Author
Journal (through CODEN or title or country)
Publication year

This table will be expanded and any proposal is welcome. The attentive reader will miss one descriptor especially important for inorganic chemists: the structure type. We omitted this with full intention. Any insider knows the great confusion existing in this field caused by the lack of standardization of structure descriptions. We intend to incorporate into the data base system the method for structure standardization developed by Parthe and Gelato (1984) just as the type classification methods designed by Hellner (1981) or Liebau (1984). Doing this we follow the line to avoid descriptors not evaluated from the data by programs. Because descriptors should be as correct as data should be.

To connect descriptors and find out the results the following commands exist:

display	show all data of a desired descriptor and their frequencies in alphabetical order on the display
find	define subset by expressions built from descriptors by logical (Boolean) operators including ranges
string	select datasets with desired character strings in the defined subset
save	store the defined subset
show	show desired categories of the defined subset on the display
print	print (or store on independent device) desired categories of the defined subset
dist	calculate atomic distances and bonding angles for specified atoms and show them on the display (or store them on independent device)
plot	draw stereopicture of the structure either automatically or after input of specifying commands
findcodn	find out journal CODEN from a journal title or from parts of it or find out full journal title for a given CODEN
help	give short information for all commands, categories or descriptors in German or English (a detailed printed manual in German and English is available)

Categories which can be displayed or printed give the next table:

Name of Compound
Mineral name and origin
Chemical formula in structured form
Title of publication
Authors
Citation with full journal title or CODEN, volume, year of
    publication, first page, last page, (issue number)
Unit cell dimensions, unit cell volume, number of formua units
Measured density

Space group symbol (Hermann-Mauguin)
Atomic parameters
  Atomic symbol
  Oxidation state
  Number of positions and Wyckoff symbol
  Atomic coordinates
  Site occupation
  Isotropic or anisotropic temperature factors
Reliability-Index
Remarks about method of measurement, specialities of the
    structure, etc.
Atomic distances
Bonding angles
Stereopicture (in many cases)
Test results
All data in standard format for further programmed use
Refcode (Cambridge Crystallographic Data File only)
Chemical Class (Cambridge Crystallographic Data File only)

In our experience the handling of the system can be learned in a few minutes. Some more time is necessary to think how to build the expressions from descriptors in the best way. This topic will be the main purpose of the data base course. Some examples may illustrate the way. The number of answers from ICSD (Release 2.8) or CCDF (Release May 84) you find in brackets.

Chemical queries:

What is the structure of mue-Carbonato-di-mue-hydroxo-bis(triamminecobalt(III)) sulfate pentahydrate?

```
  find c and o and h and n and co+3 and s and elc=6        (5)
  string mue name                                          (4)
```

What is the structure of a alloy with aluminium, iron and samarium and the stoichiometry 2:7.5:9.5?

```
  find al and fe and sm and elc=3                          (1)
```

What is the structure of Benzo-2,1-3-selenadiazole?

```
  find c and h and n and se and elc=4 and chcl=42          (9)
  string diazole name                                      (2)
```

In which binary oxides iron has several oxidation states?

```
  find O and ((fe+2 and fe+3) or (fe+2.03 to fe+2.97))
         and elc=2                                         (24)
```

What is the structure of the artificial zeolite A?

```
  display minr=zeo        (to detect the correct writing)
  find minr=zeolite                                        (222)
  string ' A ' m                                           (56)
  keep
  string artificial m                                      (42)
```

Structural queries:

Which simple cyclo-phosphates have been investigated?

```
  find p and o and met and elc=3                           (340)
  string cyclo name                                        (11)
```

Are there structures of the type ABX3 with defects in the anionic sublattice?

```
  find anx=abx3 and rem=defs                               (58)
  show p       (and inspect)
```

Which structures related to CsCl are rhombohedral distorted?

```
  find sgr=143 to 167 and anx=ax                           (39)
  show p    (and inspect coordinates by eye)
  find last and (alk or Fe)         (only relevant)        (7)
```

How often the space group P213 has been realized in case of ternary fluorine compounds?

```
  find sgr=p213 and f and elc=3                            (1)
```

Can one find a significant difference in symmetry of the carbonate group between simple carbonates of the main group and the transition elements?

```
  find c and o and (ale or alk) and elc=3
```

```
string carbonate n                                    (34)
save main
find c and o and trm and elc=3                        (94)
string carbonate m                                    (8)
dist angles from c to o
find usrn=main
dist angles from c to o
```

Analytical queries:

Which cubic crystallizing halogenides have a cell dimension of about a = 5.3 A?

```
find hal and syst=cub and cvol=144                    (13)
s e f n       (inspect)
```

Physical queries:

Which iron sulfides could be ferroelectric?

```
find sypr=pol and fe and s not o                      (10)
```

Methodical queries:

Are there very good investigations about N-H-O-hydrogen-bridging bonds by neutron diffraction at low temperatures?

```
find n and o and h and rem=tem and (rem=nds or rem=ndp)(16)
find last and d=n-h-2.0 and d=o-h-2.0                 (9)
find last and rval=0 to 0.05                          (5)
dist from h to n o dmax=2.5
```

Bibliographical queries:

Which papers N.V.Belov has published in American Mineralogist between 1960 and 1970?

```
findcodn mineralogist   (to find out CODEN)
display aut=bel      (to detect correct writing)
find codn=ammia and aut=belov and year=60 to 70       (1)
```

To find an answer to a query one should not always try to specify the query as strong as possible. Instead of this one should make a compromise between the expenditure on time to formulate the query and the number of answers which could be expected. Having done the find command the system always gives the number of true answers. Then the user can decide if he tion.

Further on it happens that the user will be astonished about the big number of answers which he receives. He didn't realize that his query would cover a much wider field than he intended. On the one hand some show commands will help him to exclude the not wanted answers. On the other hand this way has some effect of browsing in so far as the user gets more information than he had expected.

This may be demonstrated by the following example:

```
 Determined structures of sodium sulfates and their hydrates
   find na and s and o and h and elc=4 not rem=nprm           (25)
   found formulas:  Na H S O4                Na3 H (S O4)2
                    Na2 S O4 (H2 O)10        Na H S O4 (H2 O)
                    Na2 S2 O3 (H2 O)5
                    Na2 S2 O6 (H2 O)2        Na2 S4 O6 (H2 O)2
                    Na2 S (H2 O)5            Na2 S (H2 O)9
   find na and s and o and elc=3 not rem=nprm                 (15)
   found formulas:  Na2 S O3                 Na2 SO4
                    Na2 S2 O4                Na2 S2 O4
                    Na2 S4 O6
```

I think there are unexpected formulas as well as those formulas which could be expected but have not been determined until now.

Finally it will also happen that the user cannot find what he has in mind. Unsuccessful queries only based on checked data (see above) show that the data base has no relevant entry.
In all other cases the user should check the asked descriptor through the display command because e.g. names can be spelled very different.

To be user-friendly:
Access to the Inorganic Crystal Structure Data Base can happen in two ways:

1. Direct access at the host INKA. All what you need is an accession number. You will get it from Fachinformationszentrum Energie Physik Mathematik, D - 7514 Eggenstein-Leopoldshafen 2. The connection is possible by simple acoustic coupling of your telephon apparatus or much cheeper by coupling your terminal to a data network like DATEX-P (in FRG), PSS (in Britain), EURONET (in Europe), Transpac (in Canada), STN (Scientific and technical information network)(in USA, Japan). For occasional use it is the best way because you always access the newest version of the data base without any trouble for implementation etc. The common use of all three crystallographic data bases is still restricted to FRG.

2. For multiple use of the data base and if you wish to join your own evaluation programs it would be better to implement the system on your own mainframe computer. The implementation has already been checked for computers of IBM, Siemens, VAX, But be aware that it needs 30 Mbyte storage capacity. If you also want to add the Cambridge Crystallographic Data File, it needs 70 Mbyte more.

The considerable work of updating, checking and programming could only be done by the help of numerous scientists and students in Bonn and outside. Thus we got data from Clausthal, Darmstadt, Delft, Erlangen, Göttingen, Hamilton, Marburg and Parma. But as mentioned above all crystallographers can help to complete and to improve the data base by giving additional structures and corrections. Continuation has been ensured by the Fachinformationszentrum Energie Physik Mathematik if the scientific community will accept data bases by intensive use.

Literatur

Bergerhoff, G., Hundt, R., Sievers, R. and Brown, I.D. (1983). J.Chem.Inf.Comput.Sci. 23, 66-69.

Calvert, L.D. (1981). Acta Cryst. A37, C343-C344.

Hellner, E., Koch, E. and Reinhardt, A. (1981). Physik Daten - Physics Data Nr. 16-2. Karlsruhe: Fachinformationszentrum.

Liebau, F. (1984). The Crystal Chemical Classification of Silicate Anions. (in preparation)

Parthe, E. and Gelato, L.M. (1984). Acta Cryst. A40, 169-183.

THE CAMBRIDGE DATA FILE

WHAT IS IT, WHY DO WE NEED IT, AND HOW DO WE USE IT?

Robin Taylor

Crystallographic Data Centre, University Chemical Laboratory, Cambridge, UK

If we were to ask a man on the streets of New York whether he intended voting for Ronald Reagan in the forthcoming presidential election, we would probably get a clear and unequivocal answer. We would not learn the name of the next president of America. If we asked a few more people the same question, we might get a better idea of Mr. Reagan's chances, but even then our conclusions would depend on whether we were in Fifth Avenue or the Bronx. Only by asking lots of people, from all areas of the country, might we be willing to risk our lives' savings on the outcome of the election.

So it is with crystallography. I recently read a paper in which three crystal structures were described. The author wrote that his results 'confirmed' that sulphone oxygen atoms are better hydrogen-bond acceptors than carboxyl oxygens. Wrong! No analysis based on just three structures could confirm such a fact! If the author had looked at three hundred structures, he might have been entitled to draw his conclusion in such an unequivocal fashion.

The point here is that we often need to look at large numbers of crystal structures in order to obtain reliable chemical information. Unfortunately, over the years, crystallographers have tended to treat the chemical literature like a dustbin. Structures are determined for specific reasons - perhaps to identify the molecule or study its conformation - and then written up and consigned to a journal, never to be looked at again. The result of this lamentable policy is that a vast reservoir of chemical information has never been fully exploited. Just how vast is shown by Fig. 1, which plots the number of organo-carbon crystal structures published in the years 1961-80. The current total is about 40,000; by extrapolating the plot, we can estimate that 100,000 will be reached before the end of the century.

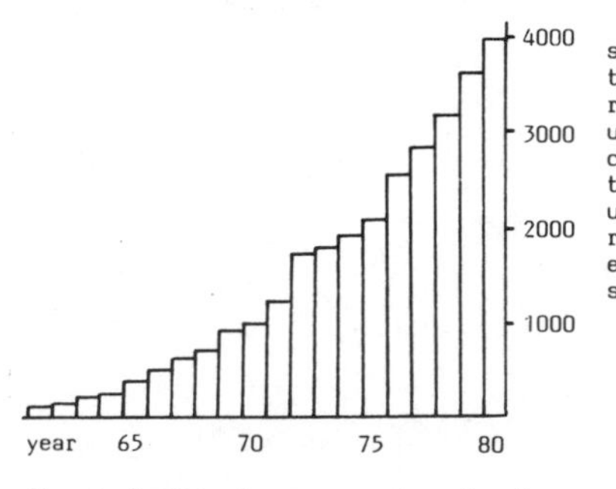

Fig. 1. Number of organo-carbon structures published between 1961-1980.

The reluctance to exploit this information is understandable and undoubtedly due to the sheer volume of data involved. This creates two specific problems, one of access, the other of analysis. For example, in order to compare hydrogen bonds involving sulphone and carboxyl oxygens, we would first need to extract from the literature those structures containing such bonds: i.e., we would have to access the relevant information. We would then need to analyse the raw crystal-structure data in order to obtain the required chemical information. Since we might be dealing with several hundred structures, this would be a non-trivial task.

In the long term, there is only one solution to each of these problems. We must use computerised databases to solve the difficulties of access, and statistical methodology to solve the difficulties of analysis. As far as databases go, crystallography is fortunate in being served by several. The Cambridge Database is discussed here, but the Protein, Metals and Inorganic Data Files should also be mentioned[1]. Turning to statistics, we find that the situation is less satisfactory: much work needs to be done in adapting standard statistical techniques to the particular needs of crystallography.

The Cambridge Structural Database

The Cambridge Structural Database contains the results of X-ray and neutron diffraction studies of organo-carbon compounds (i.e. organics, organometallics and metal complexes). Available in over 20 countries, it currently contains some 40,000 structures and grows by 10-15% each year. The database is comprehensive from 1935 onwards and is also a depository of unpublished coordinates. Each new entry to the file is checked for consistency between the reported atomic coordinates and bond lengths.

Information in the database falls into three categories (Table 1): bibliographic (BIB), connectivity (CONN) and crystallographic (DATA). This information is accessed by a system of computer programs - some examples

Table 1. Information stored in the Cambridge Database

Bibliographic (BIB)

Compound name; qualifying phrase(s) [e.g. neutron study, absolute configuration determined]; molecular formula; literature citation; chemical class(es) [e.g. 15 = benzene nitro compounds, 51 = steroids, 58 = alkaloids].

Chemical Connectivity (CONN)

Chemical structural diagrams are coded in terms of atom and bond properties. Atom properties: atom sequence number (n); element symbol (el); no. of connected non-H atoms (nca); no. of terminal H atoms (nh); net charge (ch). Bond properties: pair of atom sequence numbers (n-i,j); bond type for bond i-j (bt); [see fig. 2B,C].

Crystallographic Data (DATA)

Unit-cell parameters; space group; symmetry operators; atomic coordinates; accuracy indicators [mean estimated standard deviation of C-C bonds, R-factor]; evaluation flags [indicating presence of disorder, method of data collection, presence of errors, etc.]; comments [e.g. describing any disorder or errors in original paper].

Fig. 2. Cambridge Database program system.

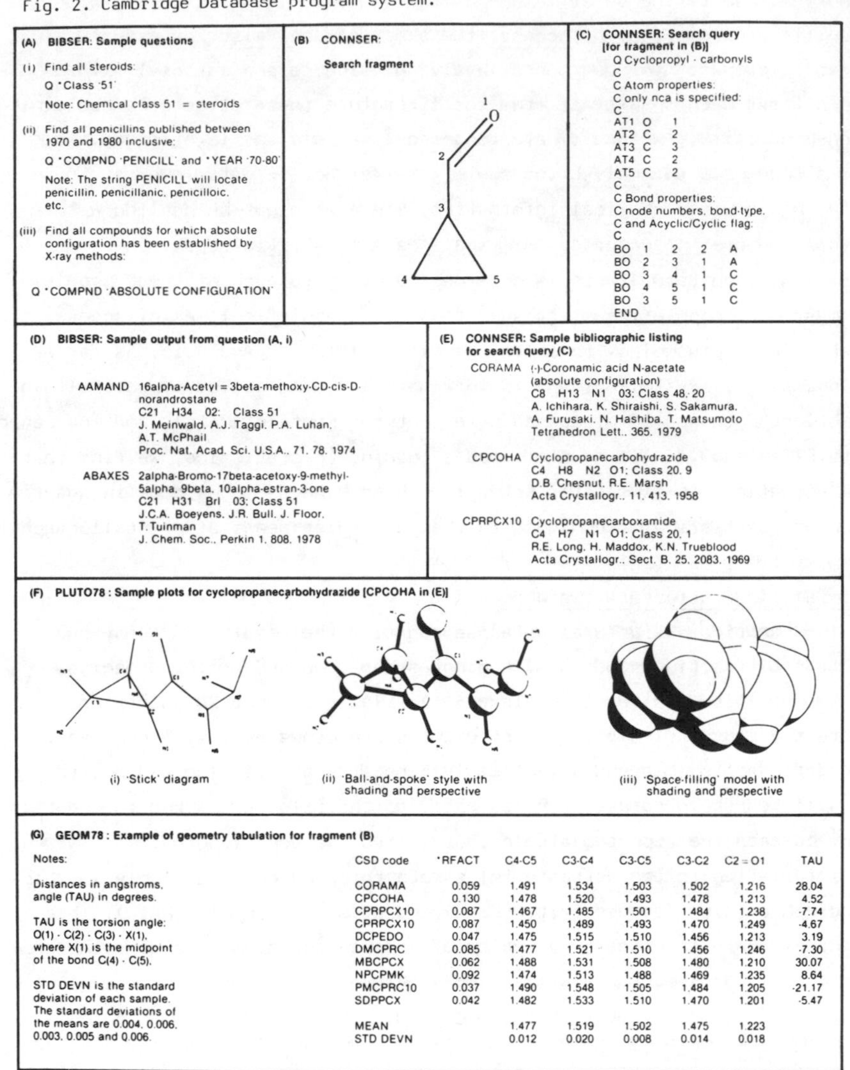

(A) BIBSER: Sample questions

(i) Find all steroids:

Q *Class 51'

Note: Chemical class 51 = steroids

(ii) Find all penicillins published between 1970 and 1980 inclusive:

Q *COMPND 'PENICILL' and *YEAR 70-80'

Note: The string PENICILL will locate penicillin, penicillanic, penicilloic, etc.

(iii) Find all compounds for which absolute configuration has been established by X-ray methods:

Q *COMPND 'ABSOLUTE CONFIGURATION'

(B) CONNSER:

Search fragment

(C) CONNSER: Search query [for fragment in (B)]

```
Q Cyclopropyl - carbonyls
C
C Atom properties:
C only nca is specified:
C
AT1 O 1
AT2 C 2
AT3 C 3
AT4 C 2
AT5 C 2
C
C Bond properties:
C node numbers, bond-type,
C and Acyclic/Cyclic flag:
C
BO 1 2 2 A
BO 2 3 1 A
BO 3 4 1 C
BO 4 5 1 C
BO 3 5 1 C
END
```

(D) BIBSER: Sample output from question (A, i)

AAMAND 16alpha-Acetyl = 3beta-methoxy-CD-cis-D-norandrostane
C21 H34 02: Class 51
J. Meinwald, A.J. Taggi, P.A. Luhan, A.T. McPhail
Proc. Nat. Acad. Sci. U.S.A., 71, 78, 1974

ABAXES 2alpha-Bromo-17beta-acetoxy-9-methyl-5alpha, 9beta, 10alpha-estran-3-one
C21 H31 Brl 03; Class 51
J.C.A. Boeyens, J.R. Bull, J. Floor, T. Tuinman
J. Chem. Soc., Perkin 1, 808, 1978

(E) CONNSER: Sample bibliographic listing for search query (C)

CORAMA (-)-Coronamic acid N-acetate (absolute configuration)
C8 H13 N1 03; Class 48, 20
A. Ichihara, K. Shiraishi, S. Sakamura, A. Furusaki, N. Hashiba, T. Matsumoto
Tetrahedron Lett., 365, 1979

CPCOHA Cyclopropanecarbohydrazide
C4 H8 N2 O1; Class 20, 9
D.B. Chesnut, R.E. Marsh
Acta Crystallogr., 11, 413, 1958

CPRPCX10 Cyclopropanecarboxamide
C4 H7 N1 O1; Class 20, 1
R.E. Long, H. Maddox, K.N. Trueblood
Acta Crystallogr., Sect. B, 25, 2083, 1969

(F) PLUTO78 : Sample plots for cyclopropanecarbohydrazide [CPCOHA in (E)]

(i) 'Stick' diagram

(ii) 'Ball-and-spoke' style with shading and perspective

(iii) 'Space-filling' model with shading and perspective

(G) GEOM78 : Example of geometry tabulation for fragment (B)

Notes:

Distances in angstroms, angle (TAU) in degrees.

TAU is the torsion angle: O(1) - C(2) - C(3) - X(1), where X(1) is the midpoint of the bond C(4) - C(5).

STD DEVN is the standard deviation of each sample. The standard deviations of the means are 0.004, 0.006, 0.003, 0.005 and 0.006.

CSD code	*RFACT	C4-C5	C3-C4	C3-C5	C3-C2	C2 = O1	TAU
CORAMA	0.059	1.491	1.534	1.503	1.502	1.216	28.04
CPCOHA	0.130	1.478	1.520	1.493	1.478	1.213	4.52
CPRPCX10	0.087	1.467	1.485	1.501	1.484	1.238	-7.74
CPRPCX10	0.087	1.450	1.489	1.493	1.470	1.249	-4.67
DCPEDO	0.047	1.475	1.515	1.510	1.456	1.213	3.19
DMCPRC	0.085	1.477	1.521	1.510	1.456	1.246	-7.30
MBCPCX	0.062	1.488	1.531	1.508	1.480	1.210	30.07
NPCPMK	0.092	1.474	1.513	1.488	1.469	1.235	8.64
PMCPRC10	0.037	1.490	1.548	1.505	1.484	1.205	-21.17
SDPPCX	0.042	1.482	1.533	1.510	1.470	1.201	-5.47
MEAN		1.477	1.519	1.502	1.475	1.223	
STD DEVN		0.012	0.020	0.008	0.014	0.018	

of program input and output are given in Fig.2. There are two main search programs, BIBSER and CONNSER. The former uses the bibliographic information fields underlined in Table 1 to locate entries on the basis of their chemical name, molecular formula, chemical class, etc. (e.g. Fig.2A). The latter is used to search for compounds containing specific chemical fragments: for example, the input in Fig.2C would be used to locate structures containing the fragment shown in Fig.2B. Output from both BIBSER and

CONNSER consists of a listing of all entries in the database which satisfy the search criteria (Fig.2D,E). The program RETRIEVE may then be used to set up a subfile containing the crystallographic data for these entries. This subfile can be processed by PLUTO, which produces plots (Fig.2F), and GEOM, which produces tabulations of geometrical parameters (Fig.2G). The DATA subfile can also be processed by the user's own programs.

Statistical Methods for Analysing Crystal-Structure Data

We see from Fig.1 that a reasonably common chemical fragment may occur in several hundred, or even several thousand, published crystal structures. Since we cannot hope to look at each structure individually, a detailed study of the fragment necessitates the use of statistical techniques.

Which statistical techniques? Well, a visit to any university library will convince the reader that the literature on statistics is huge - in other words, there is a multitude of statistical methods to choose from. It is unreasonable to suppose that only one or two of these will be relevant to crystallography; instead, we may anticipate that many statistical techniques will prove valuable, and our use of these techniques will be limited only by the speed with which we learn them and the dexterity with which we adapt them to our own particular needs. There follows an informal discussion of some of the statistical methods which have already been used in analysing crystal-structure data. The discussion is not detailed or comprehensive, but is meant to give an impression of the range of statistical methods available and the problems encountered in applying them to crystallography. For convenience, many of the examples are taken from work done in Cambridge on hydrogen bonding.

Simple Parametric Methods If we have k observations of a molecular dimension (x_i, i=1,2, ...,k), the simplest thing we can do is average them. Actually, even this is not quite as simple as it sounds because we have to remember that crystal structures differ in precision. So, when we calculate an average, do we use an unweighted mean:

$$\overline{x}_u = \Sigma x_i/k$$

or a weighted mean:

$$\overline{x}_w = [\Sigma x_i/\sigma^2(x_i)]/[\Sigma 1/\sigma^2(x_i)]$$

where $\sigma(x_i)$ is the e.s.d. of x_i? Some work has been done on this recently, and the choice between $\overline{x}_u$ and $\overline{x}_w$ is found to depend on the parameter being studied[2]. If the parameter is 'soft' (i.e. sensitive to changes in its chemical environment, such as a hydrogen-bond distance), the unweighted mean should be used; if 'hard' (i.e. insensitive to changes in its chemical environment, such as a valence-bond distance in a rigid type of molecule), the weighted mean may be preferable, although $\bar{x}_u$ is still acceptable under most circumstances.

The unweighted mean is an estimate of the true, population mean of the molecular dimension being studied. Like any estimate, it has an uncertainty (or 'standard error') associated with it, and this can be calculated as:

$$\sigma(\bar{x}_u) = [\Sigma(x_i-\bar{x}_u)^2/k(k-1)]^{\frac{1}{2}}$$

Apart from their intrinsic interest, average molecular dimensions can be useful in extracting chemical information from raw crystal-structure data. For example, theoretical calculations[3] suggest that the formation of an O-H...O bond polarises the electron density at the proton-donor group so as to increase the partial negative charge at the oxygen atom. This atom should therefore become a better hydrogen-bond acceptor. The theoretical calculations thus predict that long chains of 'cooperative' bonds (...O-H ...O-H...) should be stronger - and therefore shorter - than isolated O-H...O bonds. This was confirmed by an analysis of 24 neutron diffraction structures[4], where the mean H...O distance of cooperative bonds [1.805(9)Å] was found to be significantly shorter than the mean distance of isolated bonds [1.869(23)Å]. The power of basing an analysis on a large number of structures is well illustrated here, for how else could the theoretical prediction have been confirmed so unequivocally?

In the above example, the two means were compared by calculating the weighted difference:

$$t = (\bar{x}_1-\bar{x}_2)/[\sigma^2(\bar{x}_1)+\sigma^2(\bar{x}_2)]^{\frac{1}{2}}$$

where $\bar{x}_1$ is the mean length of the cooperative bonds, $\bar{x}_2$ is the mean of the isolated bonds, and $\sigma(\bar{x}_1)$ and $\sigma(\bar{x}_2)$ are the standard errors of the means. If there is no real difference between the lengths of cooperative and isolated bonds, the weighted difference is distributed as Student's t with about (n-2) degrees of freedom (n= total number of hydrogen bonds studied). Sometimes, however, we may wish to compare several means, and in this case the appropriate technique is analysis of variance (ANOVA). For example, Table 2A gives the mean H...O distances of various types of N-H...O=C bonds[5]. The object of the ANOVA (Table 2B) is to determine whether the differences between the means are statistically significant, i.e. too large to have reasonably arisen by chance. Essentially, the technique partitions the total variance of the observed H...O distances into two parts, the variance explained by the model (i.e. the donor/acceptor classification scheme used in Table 2A) and the residual, unexplained variance:

Table 2. H...O analysis of variance

(A) Mean H...O distances

Acceptor		Donor					
		$>$N-H	$>N^+$-H	NH_4^+	RNH_3^+	$R_2NH_2^+$	R_3NH^+
Amide/ketone/ carboxyl	Mean	1.947	1.912	1.940	1.923	1.883	1.876Å
	N(obs)	752	25	19	91	12	3
Carboxylate	Mean	1.928	1.869	1.886	1.841	1.796	1.722Å
	N(obs)	74	36	56	226	47	11

(B) Analysis of variance

Source of variance	Degrees of freedom	Sum of squares	Mean square	F
Nature of donor/acceptor	11	3.2427Å^2	0.2948Å^2	17.6
Residual (unexplained)	1340	22.4376	0.0167	
Total	1351	25.6803	0.0190	

(Approximate) variance estimates:

σ^2(total) = 0.0190, σ^2(model) = 0.0023, σ^2(residual) = 0.0167Å^2

$$\sigma^2(\text{total}) = \sigma^2(\text{model}) + \sigma^2(\text{residual})$$

In this case, a simple F-test shows that σ^2(model), though small, is significantly greater than zero - in other words, the donor/acceptor classification scheme accounts for a significant part of the H...O variance (i.e. the length of an N-H...O=C bond is dependent on the precise nature of the donor and acceptor species).

Very often in statistics, it is easy to get a significant result but difficult to work out what the result actually means. The above example is a case in point. We have a significant result, certainly, but what can we infer from it? The first thing to note is that σ^2(model) is small compared with σ^2(residual) - thus, although the length of an N-H...O=C bond _is_ dependent on the nature of the donor and acceptor groups, it is _much more_ dependent on other factors not allowed for in the analysis. Note, also,that all the results refer to the solid state. The table shows, for example, that R_3N^+-H...O bonds tend to be shorter than H_3N^+-H...O bonds _in crystals_. It does not necessarily follow that the Me_3N^+-H...O=CH_2 bond is shorter than the H_3N^+-H...O=CH_2 bond _in the gas phase_ - we could argue that the NH_4^+ ion is invariably surrounded by four acceptor species in the crystalline state, which interfere with one another and prevent the formation of very short H-bonds. In essence, the statistical analysis only tells us what the result is, not why it occurs.

Probably the most familiar statistical technique to crystallographers is regression analysis, and this, too, is fraught with difficulties of interpretation. For example, Fig.3 shows a scatterplot of N-H valence-bond and N...O hydrogen-bond distances, based on a sample of 83 N-H...O bonds observed by neutron diffraction[5]. There is obviously a marked inverse correlation between the two distance parameters, and this is confirmed by linear

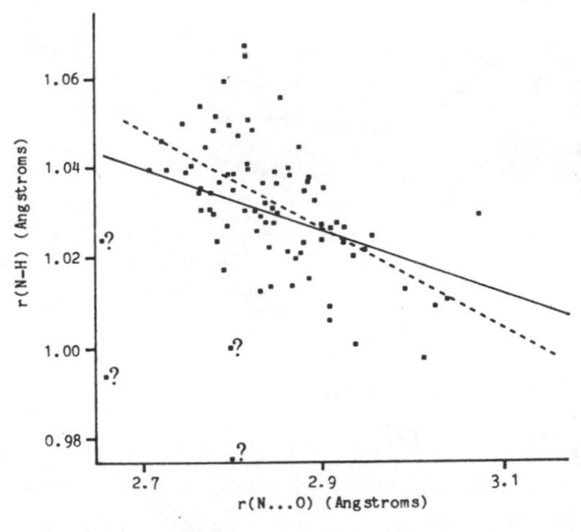

Fig. 3. Regression of N-H on N...O.

regression. The regression line is shown on the figure (solid line) and a simple test on the gradient of the line establishes that it is significantly different from zero (i.e. the value it would have if N-H and N...O were completely uncorrelated).

So what have we proved? Only that the observed N-H distances tend to get longer as the N...O distance decreases. It does not follow that the N-H bond genuinely lengthens as the hydrogen bond gets stronger, because the observed N-H distances are not corrected for thermal motion. Hydrogen bonding is known to produce a blue shift in intramolecular bending modes involving the proton-donor group, so it is possible that the thermal motion of the hydrogen atom - and hence the librational shortening of the N-H distance - gets smaller as the strength of the hydrogen bond increases. Once again, the statistical test has given us a significant result but we have to interpret that result with caution.

In the long term, the most troublesome problem in analysing crystal-structure data may be non-random sampling. Almost all common statistical techniques assume that the observations being studied are a random sample of all possible observations. Now, that is a dangerous assumption to make in crystallography because crystal structures are not determined at random! For example, many structures have been determined by neutron diffraction because they were known to contain very short, possibly symmetrical, O-H...O bonds. Consequently, an estimate of the mean length of O-H...O bonds based on all available neutron data may well be biased towards short values.

Multivariate Parametric Methods The techniques discussed so far are relatively simple, in that they look at only one or two parameters at a time. There exists a group of multivariate statistical methods which enable us to examine the distribution of several parameters simultaneously. For example, Table 3 gives the average torsion angles of twelve cytosine residues (I), together with their standard errors[6]. Suppose that we wish to

Table 3. Mean torsion angles of cytosine

Angle	Mean (S.E.)
N1-C2-N3-C4	1.4(0.8)°
C2-N3-C4-C5	-1.8(0.7)
N3-C4-C5-C6	1.3(0.8)
C4-C5-C6-N1	-0.5(0.5)
C5-C6-N1-C2	0.1(1.0)
C6-N1-C2-N3	-0.6(1.2)
C6-N1-C2-O2	-179.8(1.1)
C4-N3-C2-O2	-179.4(0.7)
C2-N3-C4-N4	178.5(0.8)
C6-C5-C4-N4	-179.0(0.8)

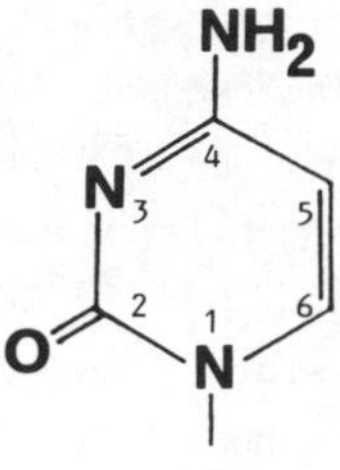

examine the hypothesis that the average geometry of the residues is planar. By and large, the mean torsion angles are consistent with this hypothesis, except, possibly, for C2-N3-C4-C5: this angle is about 2.5 standard errors away from the value expected in a planar residue (i.e. 0°). Of course, we are looking at ten torsion angles, and it is perhaps not surprising that one of them shows some deviation from its expected value. The question is: is the deviation sufficiently large to suggest that the average geometry of cytosine is not planar?

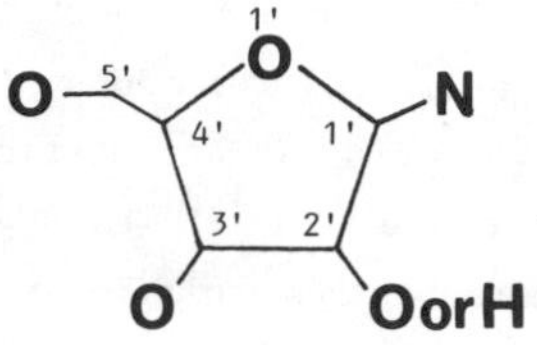

There are several ways of tackling this problem; the method actually employed was to perform a Hotelling test. This enables us to test the null hypothesis: $\bar{\tau} = \mu_0$ against the alternative hypothesis: $\bar{\tau} \neq \mu_0$, where $\bar{\tau}$ is the vector of observed mean torsion angles and μ_0 is the vector of expected mean torsion angles, assuming that the average geometry of the residue is precisely planar. The test gives a statistic which follows an F distribution, and the value obtained in this case was not significant. We can therefore conclude that the observed mean torsion angles are consistent with the hypothesis that the average geometry of the cytosine residue is planar.

Perhaps the best known of the multivariate techniques is principal-component factor analysis. The object here is to construct 'factors' - linear combinations of the parameters being studied - that explain as much as possible of the geometrical variation of the fragment under investigation. For example, a factor analysis[7] on thirteen torsion angles of the ribose (II) fragment (Table 4) showed that 61.4% of the geometrical variation of the fragment was accounted for by the first (i.e. most important) factor (F_1 in Table 4). The next two factors (F_2 and F_3) accounted for 23.3% and 15.0% of the variance, respectively. The remaining ten

Table 4. Results of ribose factor analysis

Angle	Factors		
	F_1	F_2	F_3
C1'-C2'-C3'-C4'	-1.00	0.06	-0.01
C2'-C3'-C4'-O1'	0.97	0.22	0.04
C3'-C4'-O1'-C1'	-0.66	-0.73	0.13
C4'-O1'-C1'-C2'	-0.53	0.84	-0.15
O1'-C1'-C2'-C3'	0.93	-0.35	0.06
N-C1'-O1'-C4'	-0.48	0.86	-0.13
N-C1'-C2'-C3'	0.93	-0.36	0.06
O3'-C3'-C2'-C1'	-1.00	0.04	0.01
O3'-C3'-C4'-O1'	0.97	0.23	-0.06
C5'-C4'-C3'-C2'	0.98	0.20	-0.04
C5'-C4'-O1'-C1'	-0.67	-0.72	-0.13
O5'-C5'-C4'-C3'	0.03	0.26	0.96
O5'-C5'-C4'-O1'	0.03	0.26	0.97

factors accounted for only 0.3% of the variance. The analysis therefore shows that there are just three independent factors in the conformational variability of the ribose fragment. One of them (F_3) obviously accommodates the variation in orientation of the primary alcohol group (note the large coefficients for O5'-C5'-C4'-C3' and O5'-C5'-C4'-O1'). The other two factors are more difficult to interpret, but closer analysis shows that they are connected with the phenomenon of 'pseudorotation'; i.e., F_1 and F_2 accommodate the variation in puckering of the five-membered ring.

The major difficulty with factor analysis is the need to interpret the factors, i.e. work out what they actually mean in chemical terms. This, in fact, is a general problem with multivariate methods: they are so sophisticated, and produce so much numerical output, that it is sometimes difficult to see what is going on. As a result, it is easy to misuse multivariate methods without even realising it - an instructive example is discussed by Wold and Dunn[8] (pp.10-11 of their paper).

Non-Parametric Methods All of the techniques discussed so far are parametric in nature, i.e. they make rather strong assumptions about the data being analysed. Most importantly, the data are assumed to be normally distributed, an assumption that is unlikely to be true for some molecular parameters. For example, the coefficients of skewness and kurtosis of a sample of 1,509 hydrogen-bond distances were found to be 0.80(6) and 3.49(13), respectively[5]. In a normal distribution, these quantities are zero and three, respectively. Thus, the hydrogen-bond distances are positively-skewed (i.e. very long bonds are more common then very short bonds) and exhibit positive kurtosis (i.e. the tails of the distribution are longer than those of a normal distribution with the same standard deviation). Fortunately, many parametric methods can be used quite safely provided that the coefficients of skewness and kurtosis are within about ±1 of the values expected in a normal distribution[9].

A more serious problem is posed by 'outliers' - observations that do not properly belong to the population being studied (more simply, observations that are just plain wrong). These can have a major disturbing effect on parametric tests. For example, the four question-marked points on Fig.3 are so far away from the rest of the points that there must be some doubt as to their accuracy. If the points are omitted from the regression analysis, the resulting line (broken line in Fig.3) differs appreciably in gradient from the original regression line.

A number of tests have been developed which are non-parametric in nature, i.e. make only weak assumptions about the data being analysed, and, in particular, do not assume that the observations are normally distributed. These tests are also much less sensitive to the presence of outliers than parametric tests. Non-parametric methods are available for comparing means, performing analyses of variance and calculating correlation coefficients. The essential feature of all these methods is that the original observations are replaced by ranks. For example, suppose we believed that a certain parameter, ϑ, tended to be larger than another parameter, φ. The following observations:

ϑ: 1.0,1.6,1.7,1.9,6.2,8.5; φ: 0.2,1.2,1.3,1.4,1.5,3.2

would seem to support our hypothesis, but we might be unwilling to confirm this by means of Student's t-test because the data do not appear to be normally distributed. In the non-parametric alternative to the t-test (the 'Mann-Whitney test'), the observations are ranked: thus, the smallest observation (φ=0.2) is given the rank 1, the next smallest (ϑ=1.0) is ranked 2, and so on:

ϑ: 2,7,8,9,11,12; φ: 1,3,4,5,6,10

The sum of ranks for ϑ (=49) is much larger than for φ (=29), again supporting our original hypothesis. A simple arithmetic transformation of the sums of ranks gives a statistic whose distribution is known, and which can therefore be used to calculate a significance level for the hypothesis.

When used on data that could properly be analysed by parametric methods, non-parametric techniques are wasteful of data. However, they are a viable and attractive alternative to parametric tests when the data are not normally distributed, or contain suspected outliers.

References

1. (a) Cambridge Structural Database, University Chemical Laboratory, Cambridge,UK. (b) Protein Data Bank, Brookhaven National Laboratory, Upton, NY, USA. (c) Metal Data File, NRCC, Ottawa, Canada K1A OR9. (d) Inorganic Crystal Structure Database, University of Bonn, West Germany.
2. Taylor,R. & Kennard. O. (1983) Acta Cryst.,B39, 517-525.
3. Del Bene,J. & Pople,J.A. (1973) J.Chem.Phys.,58,3605-3608.
4. Ceccarelli,C., Jeffrey,G.A. & Taylor,R. (1981).J.Mol.Struct.,70,255-271.
5. Taylor,R., Kennard,O. & Versichel,W. (1984) Acta Cryst.,B40,280-288.
6. Taylor,R. & Kennard,O. (1982) J.Am.Chem.Soc.,104,3209-3212.
7. Murray-Rust,P. & Motherwell,S. (1978) Acta Cryst.,B34,2534-2546.
8. Wold,S. & Dunn,W.J. (1983) J.Chem.Inf.Comput.Sci.,23,6-13.
9. Hamilton,W.C. "Statistics in Physical Science", New York: Ronald (1964).

PROGRAMMING ASPECTS OF CRYSTALLOGRAPHIC DATA FILES: INTERACTIVE RETRIEVAL FROM THE CAMBRIDGE DATABASE

By Pella Machin

Science and Engineering Research Council, Daresbury Laboratory, Daresbury, Warrington, WA4 4AD, U.K.

Introduction

The crystallographic data files are extremely valuable scientific resources because of the wealth of information which they contain on 3-dimensional structure. It is important that flexible retrieval, analysis and display software be available to allow scientists from a range of disciplines to access structural data quickly and easily and to use it for a variety of applications, from structure comparisons to compound identification.

Several retrieval systems are available for the different databanks. The UK Science and Engineering Research Council (SERC) Chemical Databank System (Elder, Hull, Machin and Mills, 1981) is one and it will be used here as an example for interactive retrieval and display of chemical structures from the Cambridge Structural Database (CSD). This retrieval system was originally developed by Richard Feldmann (Feldmann, 1974) in the early 1970's. It still embodies his original ideas and philosophies, but the code has been extensively re-written and enhanced over the past 9 years at SERC.

Before describing the software in detail, it is important to summarise the philosophy upon which it is based.

1. This is an Information Retrieval System, not a Database Management System (DBMS), and it is therefore concerned with providing rapid, interactive retrieval rather than more general facilities for the registration, storage and manipulation of structures. Database producers require a DBMS to organise their data, but once the database has been compiled and distributed the emphasis is entirely on information retrieval.

2. The software is interactive and relies upon indexed and inverted file techniques. This method will be described in detail and compared with alternative techniques.

3. A strength of the system (compared with many commercial packages) is that it allows retrieval by chemical structure as well as retrieval

by text and numeric values. A structure as a diagram, may be entered, matched against structures in the database, then the resulting hits may be displayed as structural diagrams.

4. The system is aimed at chemist users who are able to ask questions in chemical terms, consider the implications of the results and then refine their queries accordingly. Less interactive systems which handle searches more automatically may be simpler to operate, but are less flexible and probably give less informative results in the long run.

5. The system may be used with minimal equipment such as teletype or VDU, over wide area networks. The majority of users in the UK are university scientists and although some have access to powerful graphics terminals, most university computer centres still provide fairly limited equipment. All access is from a distance over networks and telephone lines.

Overview of the UK Interactive Chemical Databank Service

A schematic overview of the system is shown in Fig. 1. CSD contains over 40,000 organic and organometallic structures which have been solved, from 1934 to the present, by crystallographic techniques (Allen, Bellard, Brice, Cartwright, Doubleday, Higgs, Hummelink, Hummelink-Peters, Kennard, Motherwell, Rodgers and Watson, 1979). The databank has been described in detail in an earlier lecture. In summary, it provides for each reference, bibliographic information (compound name, formula, author, journal reference), chemical connectivity (essentially a description of the molecule in terms of atom types, connections and bond types), and crystallographic data (cell dimensions, space group and coordinate data). The databank is distributed on magnetic tape three times each year, in the form of one full release and two updates.

This magnetic tape is the input to the file inversion process which produces structured, indexed files. There are three stages to this process (1) extraction of each property, (2) sorting, (3) formation of indexes. The inversion of CSD is performed at Daresbury on a NAS AS/7000 mainframe and involves about 100 jobs, which are set up to run overnight and take approximately 2 hours cpu time. The resulting inverted files coupled with Feldmann-based interactive software form an on-line Chemical Retrieval System. This software is written in FORTRAN, was originally DEC10 specific but has been generalised for 32-bit word computers.

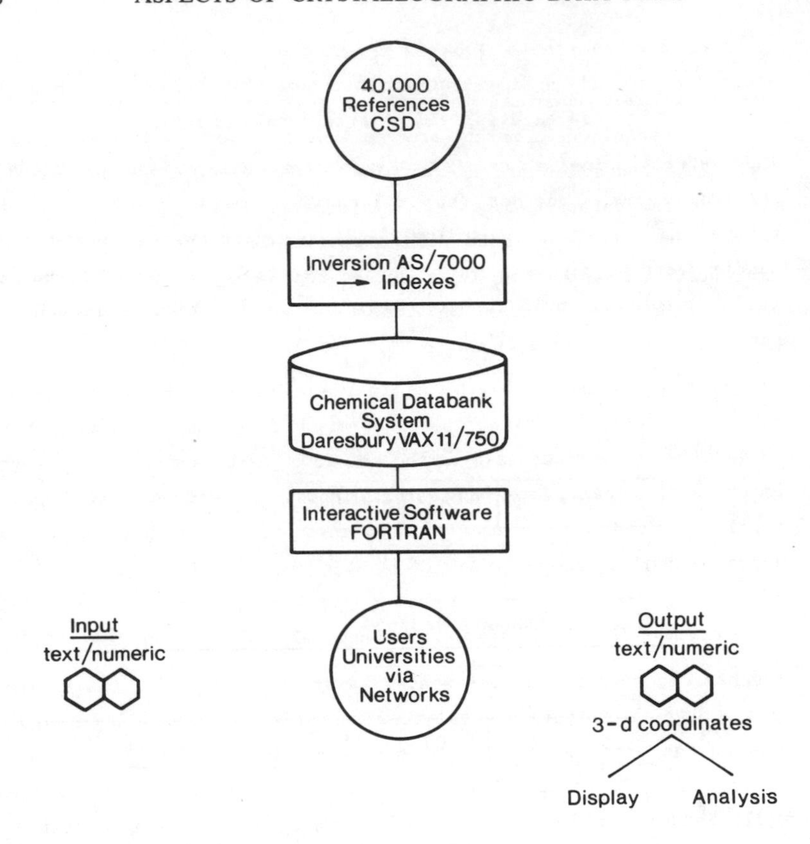

Fig. 1 Overview of the UK SERC Interactive Chemical Databank System

The Daresbury Laboratory is one of a number of UK SERC Laboratories which provide central, scientific facilities to the academic research community. The Chemical Databank System is made available as a service on a Systime 8750 with a VAX 11/750 processor at Daresbury and is used by 85 groups, mainly in universities and polytechnics, but also including some industrial companies. The current components of this system are a C^{13}nmr spectral databank, a directory of commercially available fine chemicals, and the Cambridge Structural Databank. Terminal access is via the SERC-wide area network. Lists of references and retrieved coordinate data sets may be transferred to the user's network node for printing and further display and analysis. The system may be used with ordinary teletype or VDU, though better structural diagrams can be obtained if a graphics device is available.

Inverted Files

This particular chemical information retrieval system is based on inverted file techniques, which are well-suited for flexible, on-line searching. Inverted files are hierarchical indexes which are produced by processing the raw databank material. Information for a particular category of data, usually associated with a retrieval key, is extracted from each reference, and assembled in a structured form for subsequent easy access. A schematic diagram of the inverted files associated with retrieval based on AUTHOR is shown in Fig. 2.

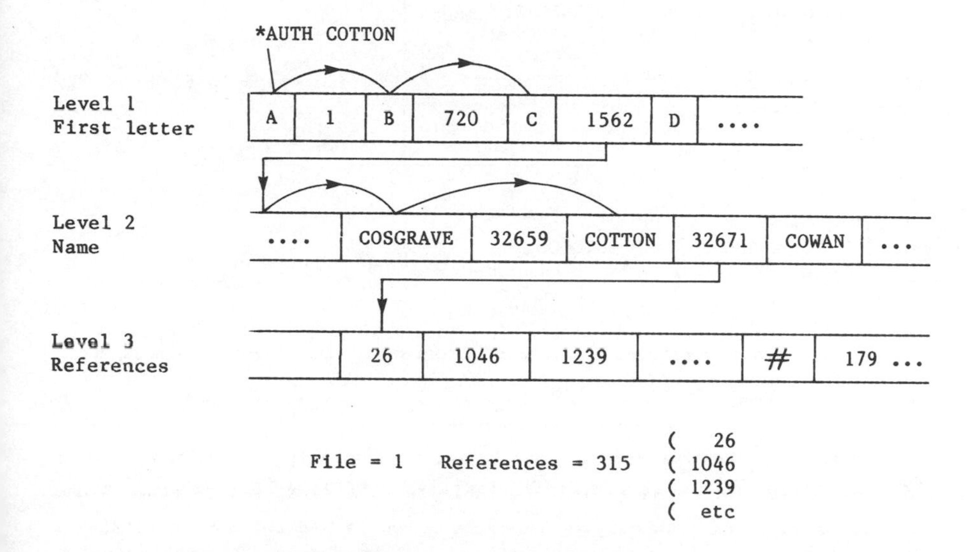

File = 1 References = 315 (26
(1046
(1239
(etc

Fig. 2 Inverted files example for author name retrieval. The retrieval progresses through two levels before a list of reference numbers of papers for the particular author is obtained. (1) The first letter of the surname, C here, is checked against the nodes of level 1, and when a match is found the pointer to file 2 is obtained. (2) This pointer is the position in file 2 where nodes containing surnames with this particular first letter begin. These nodes are checked until a match is found and the associated pointer locates the position of reference numbers in level 3.

In practice the programming associated with this simple example involves some additional niceties such as the hash representation of variable length author surnames, and an extra level (between 2 and 3) to incorporate author initials. The inverted files are constructed for direct access so retrieval incurs a few disk reads only, and response to the user's request is therefore rapid. Inverted files are set up for each keyed value including those associated with chemical connectivity searches, as well as those for bibliographic and numeric data.

Bibliographic and Numeric Data Retrieval

```
*AUTH SUNDARALINGAM                                   File = 1 References =  99
*WORD INOSINE                                           "  = 2      "     =  30
*INTER 1,2                                              "  = 3      "     =   2
*SHOW  3
37269 BEPRAP INOSINE CYCLIC-(3'-5')-MONOPHOSPHATE MONOHYDRATE
C10 H11 N4 O7 P1
H2 O1
M.SUNDARALINGAM,T.P.HAROMY,P.PRUSINER
38 1536 1982 ACTA CRYSTALLOGR., SECT.B
...
*CDIM 1 6.1, 6.2                                      File = 4 References =  227
*CDIM 5 108.4, 108.5                                    "  = 5      "     =   66
*SPGR P21                                               "  = 6      "     = 2693
*INTER 4,5,6                                            "  = 7      "     =    1
*XSHOW 7
37269 BEPRAP (C-C)ESD=1
  4 P21        6.190  13.090   9.095
R=0.034       90.000 108.430  90.000
```

Fig. 3 Example Retrieval Using Bibliographic and Numeric Data Keys

Figure 3 shows a sample search based on bibliographic and crystal data keys; similar commands are available for the retrieval of other items such as formula and a summary of these is given in Table 1. Each retrieval command is independent of the others and accesses a particular inverted file, which contains structured information of a particular category, collected from the complete database. The result of each retrieval is a temporary file of reference numbers and these temporary files may be manipulated logically to merge results or to extract references common to two or more independent searches.

Command input is straightforward and only employs character strings or numeric values. The inverted file technique ensures that searches are fast, taking typically a few seconds of elapsed time. A selection of retrieved references may be displayed at any stage and the flexible nature of the system allows the chemist user to follow up ideas quickly during an interactive session, and to modify the search strategy in the light of the results produced.

Bibliographic	Crystal Data
Author Name	Cell dimension range
Word or 4-letter word fragment in compound name	Space group
Molecular formula	Data collection method
Atom range in formula	Disordered structures
Journal reference	Error structures
- by journal name	R factor range
- by year	Calculated density range
- by exact reference	
Compound classification	
Reference number	
Cambridge CODEN	

Table 1 Bibliographic and Numeric Data Retrieval Keys

These bibliographic and data retrieval options provide useful routes to much of the information contained in CSD. However for the retrieval of chemical structures, searches based on chemical connectivity are much more powerful and comprehensive.

Connectivity Based Retrieval

A search based on chemical structure involves two main stages. First, the user must describe to the computer system the particular molecular fragment which is of interest; secondly, the actual search is performed. This involves, in effect, matching the input query structure against each of the database structures.

1. Query structure input

The chemist thinks in terms of a structural diagram to describe a particular molecule or a series of molecules. A user-friendly computer interface will, when possible, provide input facilities for such diagrams. There are a number of powerful systems available, such as those used by Chemical Abstracts, and those provided by companies such as Molecular Design and Upjohn. However, they all have the disadvantage, in the context of an academic environment, of requiring graphics devices and employing light pens or bit pads. These devices are not necessarily available to the university user, who often has no special equipment, and who expects to go along to the university computer centre or library to use any available terminal. For this kind of user it is helpful to provide a system, such as that originally devised by Richard Feldmann and still used in our present system: it employs simple text commands which can be input on any VDU, to build up and display a query structure. An example of this command language is shown in Fig. 4.

```
*NUC 65

*ABRAN 5,1
*SATOM O 10
*SBOND CD 5,10

*SATOM N 2,6,7,9
*SBOND RD 1,2,3,4,8,9
*D
```

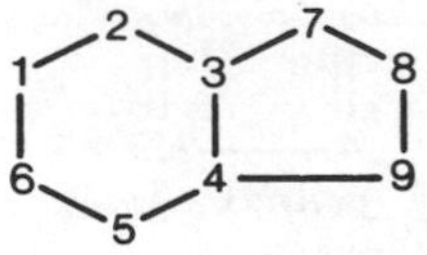

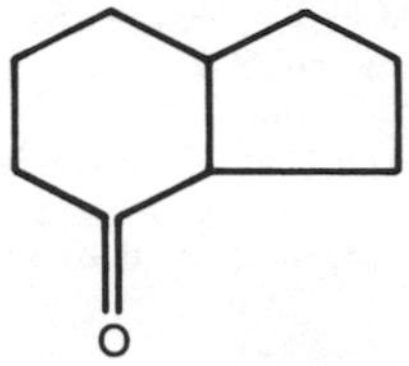

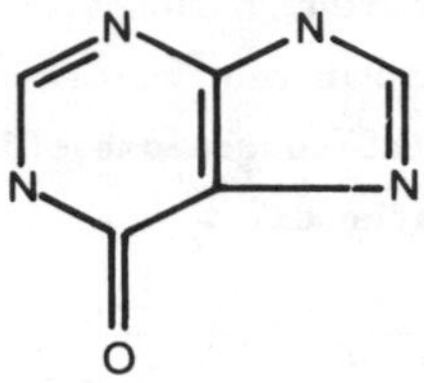

Fig. 4 Example of chemical structure input. Bond types specified here are CD = chain double and RD = ring double; the default values are 'any chain' or 'any ring'.

Device technology, however, is developing very rapidly and graphics input devices are becoming much more common and less expensive. It is a simple matter to replace the existing user-interface for chemical structure input by an alternative, without affecting the rest of the retrieval system.

2. Connection Table Representation

The connectivity of a molecule is usually thought of in graphical terms, but it is stored in the computer as a connection table. An example is shown in Fig. 5 where each row of the table represents a node, or atom, with information about its type, its connections and with additional information such as hydrogen count and charge.

Connection Table

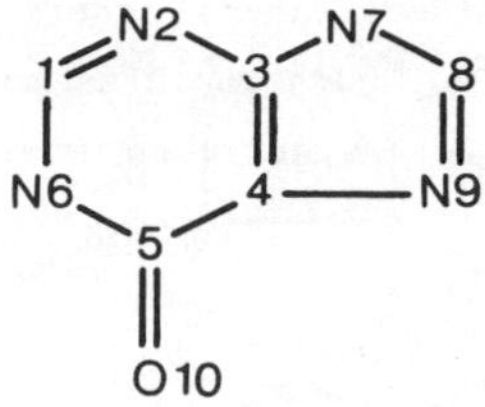

1 C 1 2(RD) 6(RS)
2 N 0 1(RD) 3(RS)
3 C 0 2(RS) 4(RD) 7(RS)
4 C 0 3(RD) 5(RS) 9(RS)
5 C 0 4(RS) 6(RS) 10(CD)
etc.

Fig. 5 Connection Table example. For each atom, information is given on which atoms are connected to it with the bond type in parenthesis. The hydrogen count is detailed in column 3.

Connection tables are provided by Cambridge for each structure in the database. They are stored for subsequent atom-by-atom matching and display purposes. These tables are usually compressed to save disk storage and to minimise I/O transfer time.

3. Screening by Fragment Probes

The aim is to match a query structure against each of the database structures in turn. For 40,000 references this would be unreasonably time consuming in an interactive program and some sort of screening is required to obtain a subset of structures which can then be used for atom-by-atom matching. Various techniques are in common use: the one used here is to provide inverted files for connectivity-based fragment retrieval purposes.

One such preliminary probe is based on ring fragments (RPROBE). The inverted file for this search has four levels and is based on, in order, ring nucleus shape, hetero atom position, hetero atom type and substituent position. Each level may be set to an exact or an embedded match so that, for example, to embed the fourth level would be to retrieve structures which not only have substituents at the ring position of the query structure, but also those which have additional substituents. Embedding the first level would retrieve structures which contain the given ring fragment in a larger ring system, as well as those in which it occurs exactly.

Another essential fragment probe is that based on atom-centred fragments (FPROBE). The important aspects of this are the type of the central atom, the types of the atoms to which it is immediately connected, bond types, and the number of occurrences of this fragment. Bond types may be set specifically (eg ring single) or generally (eg any double), and atom types may be set to a generic type embracing several atom symbols as well as to specific elements.

A combination of these probes, and miscellaneous others based on formula, may be used to reduce the original 40,000 structures to a subset of no more than a few hundred, which can be be searched exactly. An example search is shown in Fig. 6.

4. Atom-by-Atom Matching

After preliminary screening by fragment probes it will usually be necessary to perform a substructure search to obtain an exact retrieval. RPROBE does not take account of bond types, or substituent atom types and FPROBE deals with isolated atom-centred fragments, which may not be connected together in the same way as in the molecular fragment of interest. Substructure searching uses the query structure as a template and matches it atom by atom, against each database entry in the reduced set. This step is comparatively slow, but can still process approximately 500 database entries per minute.

Again it should be stressed that the philosophy of this system is to assume the user is a chemist who can consider what preliminary probes are necessary to obtain a reduced set and who can benefit from the information obtained during these intermediate stages of retrieval.

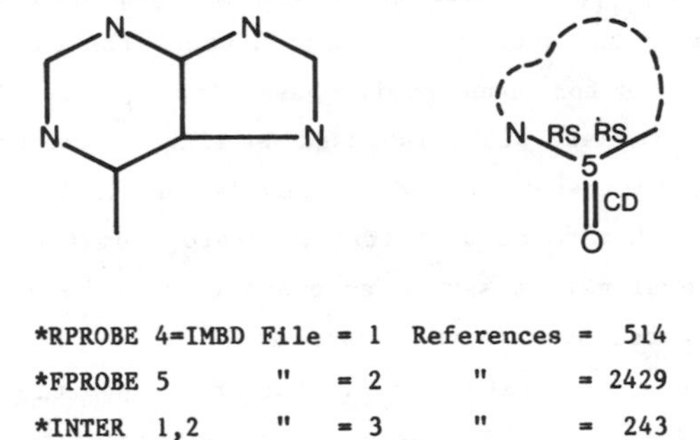

*RPROBE	4=IMBD File	= 1	References	= 514
*FPROBE	5 "	= 2	"	= 2429
*INTER	1,2 "	= 3	"	= 243
*SUBSS	3 "	= 4	"	= 36

Fig. 6 Connectivity Retrieval Example.
The ring probe (RPROBE) takes note of the 65 ring shape, the position of the nitrogen atoms, and the substituent on atom 5. A fragment probe (FPROBE) based on atom 5 finds all fragments with a 3-coordinate central carbon atom, connected to carbon by a ring bond, nitrogen by a ring bond and oxygen by a chain double bond. There are 243 references common to these two searches, 36 of which are shown to contain the full query structure after the atom-by-atom matching step.

5. Inverted Files

The fragment probes described in (3) above use inverted file techniques in just the same way as the bibliographic and numeric data commands. For example, retrieval from the four structured levels associated with the RPROBE command consists of locating the node or nodes at each level which match the query and following the associated pointer to the next level of the file. The fourth level points to a list of references of structures containing the desired ring fragment.

It is instructive to compare this inverted file fragment probe technique with a common alternative, that of serial searching using bit screens. The speed of retrieval by screen techniques is linear with the number of entries in the databank, and therefore a batch job on a large computer is often necessary. Updating such a system is however very straightforward since the bit screen for a new entry is simply added to the existing file in order to include it in subsequent searches. Inverted file systems are comparatively fast and flexible at retrieval time and can be run on-line, interactively. However, the files may take a significant amount of disk space, which must be on-line all the time, and very large databases (for example CAS with 6 million compounds) may be impossible to invert. Inverted file systems are harder to update, but are ideal for stable databases (less than 100,000 references) requiring fast, flexible retrieval.

Display of Retrieved Data

The facility to display references is an essential feature of any on-line system. A user needs to list references on the terminal, during an interactive session, and to print them to obtain hard copy. It is straightforward to display bibliographic text and numeric data and an example is shown in Fig. 3 where the SHOW and XSHOW commands are used for this purpose. It is also necessary to display structural diagrams and two methods are in common use: two dimensional chemical diagrams can be drawn, stored, then retrieved and displayed, or alternatively diagrams can be generated automatically from the stored connection tables.

Two examples of the latter approach are provided by the Daresbury interactive system. Diagrams can be produced on the most basic teletype or vdu (and are acceptable for about 80% of CSD structures) using a Feldmann-based grid display technique. A much enhanced display program (Hull 1983), requiring a graphics terminal, produces significantly improved diagrams. There are facilities within this program for showing hydrogen atoms and for compressing commonly occurring terminal groups to single symbols, as chemists often do when drawing molecules. The program generates coordinates for each atom of the molecule and uses least squares techniques to optimise bond lengths, non-bonded contacts and where possible avoids overlapping atoms or crossing bonds.

Using the Retrieved Data

CSD contains full 3-dimensional coordinate data for all entries for which structure coordinates have been published. Often retrieving references from the databank is just a preliminary step to the extraction of this coordinate data, for a small number of compounds of interest, for subsequent display and analysis. This display and analysis is best carried out locally, with some form of intelligent terminal with graphics capabilities, and therefore the availability of good network links with file transfer facilities is desirable.

There are several well-known programs which may be employed to display and manipulate structures, and to perform geometric calculations such as bond distances, angles and planes. As well as examining structures independently in this way, a chemist may want to analyse a series of molecules to compare the geometric properties of some common feature or group. Software (eg GEOM78, 1978) is available to do this in an off-line batch mode, but in many cases interactive analysis would be very advantageous, and some new software (Elder, Machin, Hull, 1984) has been written for this purpose.

CDA: Crystal Data Analyser

This software has been developed as a tool for the interactive comparison of sets of atomic coordinates from crystal structure analysis. An interactive mode of working allows the user to explore the limitations of the available data, whilst formulating a good strategy for testing hypotheses and producing displays which best illustrate the conclusions which are to be drawn from the analysis.

Three-dimensional coordinate data, such as that extracted from CSD, for a selected series of compounds, provides the raw data for the geometric analysis. The software provides a framework for matching a query structure against each of the selected compounds, by performing a substructure search based on connectivity with additional geometric constraints. The user is able to tabulate (or show as scatter plots) a wide range of geometrical parameters for comparative purposes. Structural parameters such as bond lengths and bond angles may be defined for evaluation and optionally constrained to lie within a specific range. Accuracy constraints may also be imposed.

A recent study of the geometry of the C-N-O group in some C-nitroso, isonitroso and oxime structures (Gilli, Bertolasi, Veronese, 1983) has been used here as an example. The CSD (October 1983) was searched by chemical connectivity techniques using the interactive retrieval system (Elder, Hull, Machin, Mills, 1981) and 39 structures were selected for further analysis. A CNO query structure was set up and the coordination of the C was constrained to be 3 in the subsequent matching process which resulted in 43 hits. A scatter plot of N-O versus C-N bond distances is shown in Fig. 7 and a strong correlation is evident between these two distances. An advantage of the interactive technique is that it is easy to identify the source of outliers on the plot and if necessary to eliminate them and re-plot, or alternatively to display a projection of the 3-dimensional coordinates to examine a particular structure in detail. Similar plots and tables can be obtained for N-O distance v CNO angle for example.

The software was developed for an ICL PERQ single-user minicomputer and is written in FORTRAN77. It exploits the graphics facilities offered by the bit-mapped A4 screen and uses the associated bit pad and puck for interaction with the screen. The program is menu driven and provides simultaneous display of query structure, parameter and constraint definitions, or resulting tables and plots. Consideration is being given to the possibility of making the program available in association with the Chemical Databank System on the VAX computer at Daresbury.

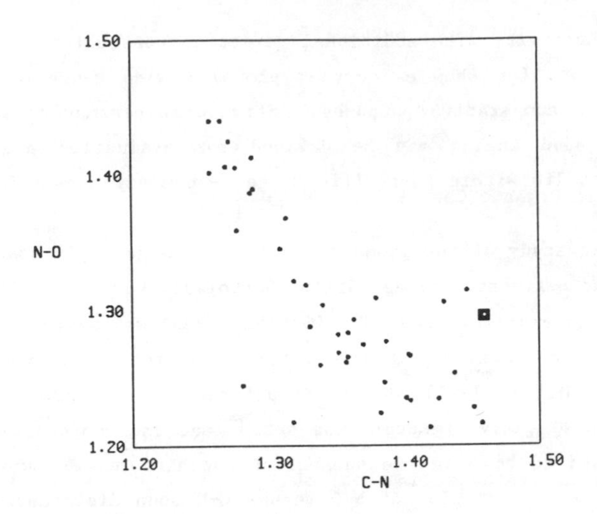

Fig. 7 Example analysis of results showing scatter plot of N-O v. C-N bond distances for some C-nitroso, isonitroso and oxime structures.

References

Allen, F.H., Bellard, S., Brice, M.D., Cartwright, B.A., Doubleday, A., Higgs, H., Hummelink, T., Hummelink-Peters, B.G., Kennard, O., Motherwell, W.D.S., Rodgers, J.R., Watson, D.G. (1979). Acta Cryst B35 2331-2339.

Elder, M., Hull, S.E., Machin, P.A., Mills, O.S. (1981) Crystal Structure Search and Retrieval, User Manual, SERC Daresbury Laboratory, Warrington, U.K.

Elder, M., Machin, P.A., Hull, S.E., (1984) CDA: An interactive program for the comparative analysis of crystal structure coordinate data. J.Mol. Graphics (in press).

Feldmann, R.J. (1974) Computer Representation and Manipulation of Chemical Information. Edited by W.T. Wipke, S.R. Heller, R.J. Feldmann and E. Hyde, p55-81 New York: John Wiley.

GEOM78 (1978) CSD User Manual.

Gilli, G., Bertolasi, V., Veronese, A.C., (1983) Acta Cryst B39 pp450-456.

Hull, S.E., (1983) 'REWARD: a program for the display of structural diagrams from connection tables' SERC Daresbury Laboratory, in preparation.

PRACTICAL APPLICATIONS OF DATA BASES

W. B. Schweizer

Laboratory for Organic Chemistry, ETH-Zurich,

8092-Zurich, Switzerland

Introduction

In the last ten to fifteen years, as x-ray analysis developed into an easy tool for analysing structures, many publications appeared where an x-ray analysis was done merely to confirm an intermediate or a final product of a reaction. This has led to an increasing number of brief reports on new X-ray structures in countless journals, most of which are not regularly read by scientists interested in crystal structures. In this reports there are hardly any comments on the structure and in many the atomic positions are also lacking. All this makes it difficult to locate and retrieve structural information from such publications.

In order to help somebody who wants to get a survey in a new field or to be informed what is going on in an established one or to avoid that the same work is done twice, it is obviously very useful to have the information in a data base. This can be done by having a set of printed collections or through a computer search and retrieval system.

Computer Data Bases

Whereas the printed book is a well known instrument that one can find in every library, data retrieval with the help of a computer is still quite new and not as frequently used as it should be. Some breakthrough has happened may be with the searches in Chemical Abstracts, where at least some primary investigations are done on a computerized data base.

The advantage in using a computer for some classical applications such as the search for a given chemical compound or for the publications of a specific author is not so enormous compared with using the printed collections. What is decisive is the possibility to combine different basic searches, each leading to many thousands of references, and to reduce them to a reasonable number of publications. Such searches can not often be coded in simple questions or keywords that can be looked up in a book. An example could be a search for molecules that have some feature in common, like a particular chemical group, a ring pattern or some combination of functional groups. Because of the complex rules of chemical nomenclature, data retrieval by names of molecules or fragments is often very difficult, whether this is done in the library or with the help of a computer. For such problems, powerful new search methods are offered by the computerized data bases. It is possible to ask for a chemical fragment in terms of a connectivity matrix or even, with a modern interactive system, as a chemical diagram drawn by a light pen or some equivalent device. Modern search programs allow people with very little experience with computers to use such a data base. A modern data retrieval program should not depend on special knowledge of the user concerning how the data are stored or how the operating system of the host computer works. A chemist who wants interatomic bond lengths in a molecule cannot be expected to check whether he has to shift the atoms in different asymmetric units or whether he has to generate part of the molecule by a symmetry operation. A modern data retrieval program should recognize the most common errors in a search question before the operating system grinds to a halt with some hardly understandable messages and possibly with loss of all the results of the previous search. Unfortunately the number of possible mistakes one can make is infinite while a program is necessary limited in those that it can detect.

Errors on Data Bases

A first source of errors is the transfer of the data from the publication to the data file. Besides the possibility of copying-errors, there are always decisions to be made about the classification of a compound and about the bond types between atoms e.g. double bond, delocalized bond, aromatic bond, etc.. Data centres can not be expected to analyse crystallographic errors, although they may be able to detect inconsistencies; for example, that bond lengths calculated from the atomic coordinates are inconsistent with published values or differ markedly from standard values. Such inconsistencies should be mentioned in a data file entry. However the people who update a data base normally do not have enough time to read publications in detail nor to check for example, the connectivity of every molecule. They usually have to accept what the authors say or rely a few built-in checks. The connectivity, for example, can be checked with a program that works with standard bonding radii. There are no problems with this method for purely organic molecules but with organometallic compounds it is a very difficult job to code the right connectivity. In the case of purely inorganic compounds the problem is still unsolved.

As an example for inconsistency in coding in the Cambridge Data File consider the following case. A search for lithium enolates with a bond between the lithium and the enolate oxygen atom will yield only some of the lithium enolates actually present in the file. For a bond according to the standard checking program the distance between the lithium and the oxygen atoms has to be less then 1.76 Å (the sum of the standard covalent bonding radii plus a tolerance factor), whereas the typical bond length for lithium enolates is around 2 Å (the sum of the ionic radii). Some compounds are coded with a Li-O bond because the authors pointed out this feature specifically, but for other compounds the bond is not coded because the automatic procedure to link bonded atoms failed to find this connection for the reasons mentioned above. This kind of inconsistency is one of the most troublesome and is not very easy to overcome. It can lead to systematic effects on the outcome of a search. Other errors like printing errors, wrong coordinates, wrong atom types, etc have a more random nature, where the missing information is scattered more evenly about a mean value.

Information obtained from Data Bases

The main use of crystallographic data bases is somewhat different from that of other chemical data files. Because there are no abstracts stored in the data bases, it is not possible to search for the characteristics, except for a few keywords, or for the physical properties of a given compound. The data stored consist of the bibliography (author names, compound name, journal), the molecular formula, the molecular connectivity and charges of the atoms, the physical data of the crystal, and the positional parameters of the atoms. Anisotropic vibration parameters are unfortunately still missing. There is some information about the measuring conditions (low temperature, neutron diffraction, visual or diffractometer data), the accuracy (R-factor,sigma range for C-C bonds) and some flags indicative of uncorrectable errors in the structure, disorder, polymeric structure, and so on.

This information can be used to check whether a given crystal structure has already been determined. It is also very convenient to obtain atomic coordinates for molecules or molecular fragments as input into direct or

Patterson search methods. Such coordinates can also be used as starting points for all kinds of energy minimisation calculations. The amount of structural information now available allows a new kind of investigations. With the help of fragment search methods the enormous amount of structural data available can be analysed to obtain new information about the properties of molecules.

Analyzing the Results of a Search

The most obvious systematic application is to find characteristic bond distances, bond angles and conformational preferences of substituents for any particular structural fragment. Such work was done before data files became available but much time was spent in the library. Because of the limited number of structures available it was only possible to look for pronounced differences in geometry. For more subtle differences the systematic and statistical errors of a few individual structures would bias the results. The structure of a molecule in a crystal is influenced by packing forces, hydrogen bonding etc. and is not necessarily identical with the structure of the isolated molecule. A large number of similar structures is required to cancel out these effects. Such an analysis has been done for the carboxylic ester group (Schweizer and Dunitz 1982) and the amide group (Chakrabarti and Dunitz 1982). Standard dimensions for bond lengths and bond angles were calculated for these groups with different types of saturated and/or unsaturated substituents. A detailed description of the estimation of average molecular dimensions from crystallographic data is given in a paper by Taylor and Kennard, 1983. This kind of analysis is straightforward and can be done today with the standard programs that are supplied with the data bases. Some caution is necessary in selecting the data, as will be discussed in the next chapter. The standard value is a single number for a given geometrical parameter with an estimation of its standard deviation. The result can also be presented as a histogram to show the actual distribution of the data points.

Another way to analyze the data is to look for correlations among relevant structural parameters for many molecules with common functional groups (Buergi 1973; Buergi, Dunitz and Shefter 1973; Murray-Rust, Buergi and Dunitz 1975). Each fragment in a particular environment provides a sample point in a many-dimensional space, one dimension for each variable structure parameter. The scatter- plots obtained from such structural correlations mostly show a non uniform distribution of the sample points. For example, the distribution of the two dimensional plot of the distance and the angle of a nucleophile close to the sulfur atom of divalent sulfides X-S-Y shows a concentration of points opposite the S-X or S-Y bonds (Rosenfield, Parthasarathy and Dunitz 1977). If one assumes that the sample points tend to concentrate in low lying regions of the potential energy surface the observed distribution can be interpreted in terms of a reaction path for nucleophilic attack on sulfur.

An example for a indirect observation of a fairly detailed reaction path is the stereoisomerization of triphenylphosphineoxide (Bye, Schweizer and Dunitz, 1982). The three phenyl rings were treated as rigid rotors and the three torsion angles of each observed conformation were taken as a sample point in a three-dimensional space. The distribution of these points show a cluster corresponding to the symmetric propellor shape arangement of the three phenyl rings with torsion angles of about 40 degrees. There is a thinning-out of points in the region of the mirror-symmetric structure with angles 90, 10, -10 degrees. This region can be identified as the transition state for the stereoisomerization.

It is sometimes not easy to decide whether the variation of the geometry of a molecular fragment in different crystals is due to structural variation or experimental errors. Factor analysis provides an automatic way to reduce the dimensionality of such problems and to identify the main factors responsible for the observed variance of the molecular geometry (Murray- Rust, Bland, 1978). For example, factor analysis applied to the torsion angles of the ribose ring of nucleosides showed that practically only two independent factors are responsible for the variability of the ring conformation (Murray-Rust, Motherwell, 1978) in agreement with the conclusion based on counting the degrees of freedom for the out-of-plane deformations of five membered rings.

Errors in the Analysis

The easier it is to make such systematic searches, the greater the chance to make mistakes in analysing the data. The automatisation of the whole procedure leads to dangerous consequences in using data files uncritically and in bypassing the original papers in the journals. The accuracy of the data is not improved by a computer program even if it prints the result to five decimal places. The molecular parameters associated with the compounds on a data file are affected by all the possible experimental errors in crystallography. The data file contains an assortment of molecules measured under many different conditions at various temperatures with different vibrational motion, and affected by disorder, pseudosymmetry, wrong cell dimensions,etc.. Parameters from constrained refinements can also be misleading if they are not recognized as such.

The geometrical parameters that are most sensitive to such errors are the interatomic distances. Special care has to be taken in calculating mean bond lengths. Only the best structures with no disorder and no heavy atoms should be used for the analysis. There is still the problem that the bonds tend to be too short because of thermal motion. Bond angles and torsion angles are somewhat less sensitive and more structures can be used for those analyses. Disorder or strong thermal motion in a molecular fragment can have a systematic effect on several parameters, and lead to wrong correlations, e.g. a bond shortening coupled with angle opening. When inspecting results of such analyses it is important to check for outliers and refer to the original literature to verify that they are not caused by errors. A group of very similar molecules or the multiple redetermination of the same molecule can provide a clustering of sample points that is not truly representative.

With some critical use of the data files it is possible to carry out analyses of molecular geometry in a short time. Data bases can be used to increase out knowledge in many fields in crystallography and chemistry.

Literature

Allen,F.H., Kennard,O. and Taylor,R. (1982). Acc.Chem.Res. 16, 146-152.

Buergi,H.B. (1973). Inorg. Chem. 12,2321-2325.

Buergi,H.B., Dunitz,J.D. and Shefter,E. (1973). J. Am. Chem. Soc. 95, 5065-5067.

Bye,E., Schweizer,W.B. and Dunitz,J.D. (1982). J. Am. Chem. Soc. 104, 5893-5898.

Chakrabarti,P. and Dunitz,J.D. (1982). Helv. Chim. Acta, 65, 1555-1562.

Murray-Rust,P. and Bland,R. (1978). Acta Cryst. B34, 2527-2533.

Murray-Rust,P.,Buergi,H.B. and Dunitz,J.D. (1975). J. Am. Chem. Soc. 97, 921-922.

Murray-Rust,P. and Motherwell,S. (1978). Acta Cryst. B34, 2534-2546.

Rosenfield,R.E., Parthasarathy,R. and Dunitz,J.D. (1977). J. Am. Chem. Soc. 99, 4860-4862.

Schweizer,W.B. and Dunitz,J.D. (1982). Helv. Chim. Acta, 65, 1547-1554.

Taylor,R., Kennard,O. (1983). Acta Cryst. B39, 517-525.

Program systems for maxi-, mini-, and micro-computers

SURVEY OF CRYSTALLOGRAPHIC PROGRAM PACKAGES

Karel Huml
Institute of Macromolecular Chemistry, Czechoslovak Academy of Sciences,
Praha 6, Czechoslovakia

1. Introduction

Crystal structure analysis as a laboratory tool needs not only automated hardware like on-line diffractometers but also "automated" software which includes packages of scientific programs. Generally, we shall call program package (program system) a set of programs, each of which is capable of performing a different calculation and between which results are exchanged via common data base (Hall & Stewart, 1980). Program package may be a small group of specialized routines as well as a set of programs capable of solving a rather broad range of problems. But always there are several desirable features of a good program package to be fulfilled (Stewart, 1978), particularly:

(1) Documentation in the form of a handbook aimed at the novice user.
(2) Simple input and understandable output which follow IUCr and JCPDS recommendations as to their formats (Calvert et al, 1980; Young et al, 1982; Brown, 1983).
(3) Generality in the range of structure sizes, space groups and procedural variations. Package should be machine independent in order to be portable.
(4) FORTRAN language is usually expected. The use of preprocessors in connection with FORTRAN has been described by Hall (1983). An opposite policy, i.e. the use of a simple subset of FORTRAN statements, was successfully applied by Sheldrick (1981). Advantages and drawbacks of other languages have been discussed by Rollett (1970) and Stewart (1982).
(5) Reliability. Standard test data (Ahmed et al, 1972) are extensively employed. Feedback from the users and newsletters written by authors are very useful.

Unfortunately, these points often demand opposite strategies in the system design. An experienced programmer-crystallographer has to find the optimal solution. The objective of this lecture is only a short introduction into the literature on program packages in a limited area of crystallography. It does not substitute the IUCr World List of Crystallographic Computer Programs; neither is it an advertisement or criticism of any crystallographic program. All references mentioned here are only examples.

2. Powder Crystallography

2.1. Phase Analysis

Some of the first programs for the identification of phases from their powder diffraction patterns were written by Frevel (1965) and Nichols (1966). Further development led to more efficient search-match methods using the full JCPDS Powder Diffraction File (PDF) (Johnson & Vand, 1967) or specially oriented subfiles. For details on PDF see Calvert et al (1980). The introduction of a new generation of mini- and micro-computers allowed the use of an in-house inexpensive computing facility (Hare et al, 1982). A comparison of several programs was given by Nichols & Johnson (1980) and results of the Round Robin test by Jenkins & Hubbard (1979). The final goal of these methods is a reliable quantitative powder diffraction phase analysis (Fiala & Melichar, 1984).

Examples of Search-Match Program Packages

Reference	Package name	Full PDF/ subfile only	Minicomp. version available	Identif. or quant. analysis
Burova et al(1977)		Sub.	No	Quant.
Edmonds(1980)	ZRD	Sub.	Yes	Ident.
Huang & Parrish(1982)		Full	Yes	Ident.
Johnson(1977,1981)	PDSM	Full	Yes	Ident.
Lin Tian-Hui et al(1983)		Sub.	Yes	Quant.
Nichols & Johnson(1980)	SEARCH	Full		
O'Connor & Bagliani(1976)	WAIT	Full	Yes	Ident.
Schreiner et al(1982)	SANDMAN	Full	Yes	Ident.
Weiss et al(1983)	XQPA			

2.2. Indexing and Lattice Parameters

Unit cell parameters received as a result of powder indexing procedure are an excellent tool for phase identification. The corresponding Crystal Data base is twice as large as JCPDS Powder Diffraction File. However, indexing methods with some exceptions of the highest symmetry cases (Paszkowicz & Rozbiewska, 1983) are usually employed for the single solid phase and not for mixtures. Thus, methods described in the previous chapter will be dominant in the phase identification procedures, especially when a quantitative analysis is required. However, indexing of powder diagrams has gained added interest in recent years from the increased application of powder diffraction to the refinement or complete determination of crystal structures.

Shirley (1980, 1983) has shown that some indexing methods search for solution mainly in index-space, by varying hkl indices, while others search mainly in parameter-space, by varying cell parameters. Programs may also by classified according to whether they use the deductive or exhaustive approach. The speed achieved in the case of deductive methods is paid by rigour. Exhaustive methods are more rigorous but slower. For this reason, high effective indexing program packages contain more than one of these methods. They are applied subsequently with respect to the symmetry restrictions used (Shirley, 1978).

Examples of Indexing Program Packages
(+ means "and all higher systems")

Reference	Package name	Deductive or exhaustive	Space	Crystal system
Chigarov et al(1981)				
Kohlbeck & Hörl (1976,1978)	TMO	Semi-exhaust.	Index	Monocl.+
Louër & Vargas(1982)	DICVOL	Exhaustive	Param.	Monocl.+
Paszkowicz Rozbiewska(1983)				Higher
Shirley & Louër(1978)	LZON	Semi-exhaust.	Param.	Any
Smith & Kahara(1975)	QTEST	Deductive	Index	Monocl.
Taupin(1973)	POWDER	Exhaustive	Index	Any
Visser(1969)	ITO,FZON	Deductive	Param.	Any
Werner(1964,1976)	TREOR	Semi-exhaust.	Index	Monocl.+

2.3. Structure Refinement with Powder Data

Rietveld (1967,1969) developed a full-pattern-refinement method of crystal structure directly from the neutron powder diffraction pattern, without first extracting the structure factors. A review for neutrons was published by Cheetham & Taylor (1977), and for X-rays by Young (1980). Albinati & Willis (1982) gave also a survey of Rietveld methods, inclusive those for the simultaneous refinement of two or more phases described by Worlton et al (1976), Werner et al (1979), and Wiles & Young (1981).

An alternative two-step procedure was used for neutron studies (Will et al, 1965) and for X-rays (Taupin, 1973; Mortier & Costenoble, 1973; Huang & Parrish, 1975; Hecq, 1981; Will et al, 1983). First integrated intensities of the individual reflections are determined from the observed pattern, taking into account the instrument function. In the second step, the structure parameters are refined on integrated intensities in a manner analogous to the single crystal structure determination. Simultaneous refinement of two phases by this method was described by Werner (1980).

Similarities between these two methods were discussed by Prince (1981). For nomenclature see Young et al (1982).

An inversion task: calculation of a powder scattergram to the known structure is possible with the programs LAZY PULVERIX (Yvon et al, 1977) or FINAX (Hoveestreydt, 1983).

Examples of Structure Refinement Packages with Powder Data
(R-modified Rietveld program, I-integr. intensities two-step methods)

Reference	Package name	Method	Radiation
Bacon et al(1979)	EDINP	R	n, 1.38 Å
Bearlocher & Hepp(1981)	BH		X(n), α_1
Cooper et al(1981)	SCRAP	I	n,
von Dreele et al(1982)	RVD	R	n, (time of flihgt)
Hewat(1973)	RH	R	n,
Immirzi(1980)	PREFIN		X(n), $\alpha_1 + \alpha_2$
Khattak & Cox(1977)	RKC	R	X, β
Malmros & Thomas(1977)	RMT	R	X, α_1
Matthewman et al(1982)	CCSL	R	X, β
Mortier(1980)		I	X, $\alpha_1 + \alpha_2$
Pawley et al(1977)	EDINP		n,
Prince(1980)	RP	R	n,
Smith et al(1979)	POWDER	R	n,
Toray & Marumo(1980)	PFLS		X, $\alpha_1 + \alpha_2$
Werner et al(1979)	RWSMT	R	X, α_1
Wiles & Young(1981)	DBW2.9		X(n), $\alpha_1 + \alpha_2$ or α_1
Will(1979)	POWLS	I	X(n), $\alpha_1 + \alpha_2$
Young et al(1977)	RHVM	R	X, $\alpha_1 + \alpha_2$

2.4. General Purpose Powder Packages

In this category should be mentioned packages like POWDER (Osgood & Snyder, 1980), PODA (Coyle, 1981), NBS AIDS80 (Hubbard et al, 1982) designed to solve different tasks in powder analysis. Also, the X-ray equipment producers offer powder program packages, e.g. DIFRAC 300/11-Siemens, PW 1700-Philips, STADI 2/PL-Stoe and NICOLET system (see the company manuals).

3. Single Crystal Structure Analysis

3.1. Patterson Methods

The basis for the automatic interpretation of the Patterson maps was laid by Harker (1936), Wrinch (1939), and others. Related methods comprise structure-solving techniques based on the interpretation of the Patterson function in terms of some a priori known or assumed structural or (non-crystallographic) symmetry feature. Search for the Patterson function can be performed in the vector space or in the reciprocal space (Nordman, 1980). The formalism of the two types of methods is in principle inter-convertible (Huber, 1970).

Examples of Patterson Methods Packages

Reference	Package name	Place of origin	Space	Note
Crowther & Blow(1967)		Cambridge	Rec.	
Egert(1983)		Göttingen		Combined with direct methods
Furey(1983)	AP	Pittsburgh		Array processor
Harada et al(1981)		Paris	Rec.	
Hornstra(1970)		Eindhoven	Vector	
Huber(1970)		München	Vector	
Kutschabsky & Reck (1976)		E.Berlin	Vector	
Langs(1975)		Buffalo	Rec.	
Lenstra & Schoone (1973)		Utrecht	Vector	
Luger & Fuchs(1983)	IMPAS	W.Berlin	Vector	
Nordman(1980)		Michigan	Vector	
Simonov(1982)		Moscow	Vector	
Tollin(1970)		Dundee	Rec.	

3.2. Direct Methods

The real success for direct methods came when Karle & Karle (1963) started to systematize the symbolic addition method (Schenk, 1980). A multisolution approach to the phase determination using tangent formula given by Karle & Hauptman (1956) appeared latter. For a survey see Woolfson (1983). A general review on direct methods was published by Gilmore (1982). Application of direct methods for macromolecules was described by Karle (1983).

Examples of Direct Methods Packages

Reference	Package name	N-phase (sem) invariant Minicomp. version available	Note
Beurskens (1983)	DIRDIF		Symbolic add.
Giacovazzo & Cascarano (1981)	SIR	1,2,3,4	Multisolution
Gilmore(1984)	MITHRIL (NATO)	1,2,3,4,5	Package of known direct methods
Kiers & Schenk(1983)	SIMPEL83	3,4,5,yes	Symbolic add.
Langs & Hauptman(1978)	QTAN	3,4,yes	Magic integers
Riche(1982)	DEVIN	3,4,5	Symbolic add.
Woolfson(1983)	MULTAN80	3	Phase permut.
	MAGEX80	3	Magic integers
	YZARC80	3,yes	Random phases lin. eq.˘s refinement
	RANTAN81	3	Random phases

3.3. Structure refinement

Methods of the least-squares (LS) analysis for the refinement of the crystal structures were introduced by Hughes (1941). The increasing power of computers gave a chance to use the LS methods even for protein single crystals. But the usually low quality and incompleteness of diffraction data of macromolecules raised additional problems. It was also clear that it is unlikely that a conventional method of LS refinement will be used routinely for macromolecular structures, since the compution time required is proportional to NM^2, where N is the number of reflections and M the number of parameters in refinement. Some of the problems were solved by an application of LS procedures originally designed for small molecules which respect rigid-body groups of atoms and other subsidiary conditions (Scheringer, 1963; Waser, 1963). A significant contribution to the LS technique was the FFT algorithm (Agarwal, 1978), since the computation time required here is approximately proportional to NlogN. For further details see Isaacs (1982,1983) and Proc. of the Daresbury Study Weekend (Machin et al, 1981), where programs are discussed for the refinement of macromolecules defined in the real space (Diamond, 1971) and also those working in the reciprocal space.

There are other fields of application of LS refinement methods in crystallography with their own program packages, e.g. the detail study of electron density distribution (Coppens et al, 1979).

Examples of Structure Refinement Packages

Reference	Package name	Mole-cules	Constraints Restraints	Note
Coppens et al(1979)	RADIEL	Small		Ref.of extinction, struct. end el. density param's
Diamond(1971)		Macro	Con.	Real-space ref.
Dodson et al(1976)	MODELFIT	Macro	Res.	
Finger & Prince(1975)	RFINE4	Small		Robust/Resistent
Hoard & Nordman(1979)	FGLS	Macro	Con.-Res.	Gauss-Siedel LS
Jack & Levitt(1978)	EREF	Macro	Res.	Energy and R factor rec.-space ref.
Konnert & Hendrickson (1980)	RLS	Macro	Res.	Rec.-space ref.
Sussman(1983)	CORELS	Macro	Con.-Res.	Rec.-space ref.
Takano(1977)	RLSP	Macro		Real-space ref.
Wlodawer & Hendrickson(1982)	PROLSQ	Macro	Res.	X-ray and neutron joint ref.
Zucker et al(1981)	PROMETHEUS	Small		Disordered struc.

3.4. Graphics

As early as in 1965 Johnson wrote an ORTEP program which is still very popular in crystallography. A year later Levinthal published a procedure for folding macromolecules using a dynamic interactive system. Both systems work with a line-drawing equipment like the pen-and-ink plotter, CRT stroke vector display or direct-view storage tube. According to the hardware used we talked about vector graphics or calligraphy. The stroke vector CRT display has the fastest response for dynamic interactive line graphics, but it is expensive and improper for a continuous surface representation.

Other group of programs is based on raster graphics using plasma panel, printer plotter or TV monitor with a refresh buffer as a display. It is relatively cheap and excellent for space-filling model drawings, inclusive shading of the smooth surfaces. But lines have a stair-step appearance, especially when the TV screeen is divided into a small number of dots (pixels).

For a short review on graphics, especially raster technique, see Johnson (1980). Hardware for vector graphics is described by Barry & McAlister (1982) and software by Diamond (1980). Comparison of both methods is given by Connolly (1983). General principles are given by Newman & Sproul (1979).

Examples of Graphics Packages

Reference	Package name	Hardware	Mode of operation
Vector Graphics:			
Bandel & Sussman(1983)	PLORTEP	Plotter	Batch
Brandenburg et al(1981)	GUIDE	E&S PS2,CDC170 PDPLL/34	Dyn.inter.
Busetta et al(1983)	DOCKER	E&S PS2 PDP11/60	Dyn.inter.
Chabot et al(1981)			Dyn.inter.
Connolly(1982)	MS		Dyn.inter.
Diamond(1983)	BILDER	E&S PS1,PDP11/50	Dyn.inter.
Jones(1982)	FRODO	VC3404,PDP11/40 Siemens 4004	Dyn.inter.
Ksenofontova et al (1978)	GRAFIKA1	Plotter HP-2100,M-220	Batch
Langridge et al(1981)		E&S PS2,PDP11/40 CDC7600	Dyn.inter.
Lejeune et al(1983)	WAND7200	MEGATEX 7210 PDP11/60	Dyn.inter.
Motherwell & Clegg(1978)	PLUTO	Plotter	Batch

O´Donnell & Olson(1982)	GRAMPS	E&S MPS VAX11/780	Dyn.inter.
Sielecki et al(1982)	M3	MMS-X	Dyn.inter.
Spek(1982)	EUCLID82	Plotter,CYBER175	Batch
Swanson et al(1982)	PACK	VG3,PDP11/40	Inter.
Wright(1982)	GRIP75	VT11,PDP11/40	Dyn.inter.
Raster Graphics:			
Feldman et al(1978)			Step-wise
Johnson(1980)	ORCIG	GENISCO GTC-3000 ECLIPSE S/130	Batch Step-wise
Meyer(1970)	DISPLAY	XDS Sigma 7	Step-wise
Norrestam(1981)	VANDER		
Richardson(1978)	CRYSX	Elst.printer-plotter,PDP11 CDC 6600	Batch

3.5. Complete Single Crystal Packages

Packages with full range of programs necessary for the single crystal structure determination have been published since the early sixties, e.g. a system for KDF 9 computers (Rollett, 1970). Today more than 30 complete single crystal packages are currently referred in literature. The computations provided by these packages are: (1) Literature and data base searching, (2) Diffractometer and film scanner control, (3) Unit cell and space group determination, (4) Data reduction, incl. 1/Lp calculation, absorption and extinction corrections, etc, (5) Phase problem solution (Patterson and direct methods), (6) Structure refinement, (7) Thermal motion analysis, (8) Molecular geometry calculation, (9) Error analysis and hypotheses testing, (10) Graphics and preparation of tables and manuscript, (11) Special programs concerning deformation electron density, protein crystallography, etc.

Statistics on Acta Crystallographica C39 (1983) shows that approximately 35% of authors used SHELX/SHELXTL systems, 35% some of the Oak Ridge programs, and 16% XRAY/XTAL packages. General review on packages was given by Stewart (1982). Minipackages have been discussed by Gabe (1980).

Examples of Complete Single Crystal Packages

Reference	Package name	Computer	Note
Ahmed(1978)	NRC	IBM/360	
Andrianov(1978)	RÖNTGEN75	BESM-6	
Burzlaff et al(1978)	CRYSTAN	PDP11/45	Philips Software
Busing et al(1963,1964, 1981); Johnson(1976)	ORNL		ORGLS,ORFFE,WMIN ORTEP II
Davies et al(1970)	COSMOS		
Finger & Prince(1975)	RFINE 4		
Frenz(1983)	SDP-PLUS	PDP11	Nonius Software
Gabe & Lee(1981)	NRC PDP-11	PDP11	
Gilmore et al(1981)	GX	32bit mini	Fully interactive
Goodacre & Lee(1971)	LUX		NRC+OR+local pgms
Ipatova et al(1974)		MINSK22/32	
Kennard(1984)	APPLECRYST83	APPLE II	Sci.and teach.pgms
Meyer et al(1974)	CRYSNET	CDC7600	
Norrestam(1982)	XTAPL		APL programs
Sakurai & Kobayashi(1979)	UNICS III		
Sheldrick(1979)	SHELXTL	Mini	Nicolet Software
Sheldrick(1976)	SHELX76	Mainframe	
Sparks(1978)	E-XTL	Mini	Syntex Software
Stewart et al(1976)	XRAY76	Mainframe	
Stewart(1982)	XTAL	Mainframe	Preprocessor
Tovbis & Schedrin(1970)	KRISTAL		
Watkin(1982)	CRYSTALS	Mainframe	32 bit architect.
White(1984)		WS-150	16/32 bit micro

4. Future of Program Packages

Applied crystallography shows convergency to the usage of a limited number of specialized program packages. Examples are MULTAN and ORTEP referred respectively by 40 and 29% of authors in Acta Crystallographica, C39 (1983). Popularity of high-efficient and user-oriented minipackages, like SHELXTL, is growing with increasing quality of low-cost minicomputers. However, large program systems, like XTAL, will be of use in connection with mainframe computers owing to their generality, especially when a broad spectrum of problems concerning small molecules and also macromolecules are to be solved in one laboratory. Approximately 50% of authors are still using a mixture of different packages, partially because they are not fully satisfied with the existing ones. It is challenging for those crystallographers who are interested in writing new software.

The author is grateful to all colleagues-crystallographers and X-ray equipment manufacturers who kindly cooperated during the preparation of this paper.

References

Abbreviations of the IUCr Summer Schools:

(1970) Ottawa = Crystallographic Computing, Ed. F.R.Ahmed et al., Copenhagen: Munksgaard.
(1976) Praha = Crystallographic Computing Techniques, Ed. F.R.Ahmed et al., Copenhagen: Munksgaard.
(1978) Twente = Computing in Crystallography, Ed. H.Schenk et al., Delft: Delft University Press.
(1980) Bangalore = Computing in Crystallography, Ed. R.Diamond et al., Bangalore: The Indian Acad. of Sci.
(1982) Ottawa = Computational Crystallography, Ed. D.Sayre, Oxford: Clarendon Press.
(1983) Kyoto = Internat.Summer School on Cryst.Computing, Kyoto: IUCr Comp.Commission and The Cryst.Soc. of Japan

Agarwal, R.C. (1978). Acta Cryst. A34, 791-809.
Ahmed, F.R. (1978). Twente. 17-29.
Ahmed, F.R., Cruickshank, D.W.J., Larson, A.C. and Stewart, J.M. (1972). Acta Cryst. A28, 365-393.
Albinati, A. and Willis, B.T.M. (1982). J.Appl.Cryst. 15, 361-374.
Andrianov, V.I. (1978). Twente. 45-51.
Bacon, G.E., Lisher, E.J. and Pawley, G.S. (1979). Acta Cryst. B35, 1400-1403.
Bandel, G. and Sussman, J.L. (1983). J.Appl.Cryst. 16, 650-651.
Barry, C.D. and McAlister, J.P. (1982). Ottawa. 274-285.
Bearlocher, Ch. and Hepp, A. (1981). Acta Cryst. A37, C-282.
Beurskens, P.T. (1983). DIRDIF, ORIENT, TRADIR, Univ. of Nijmegen.
Brandenburg, N.P., Dempsey, S., Dijkstra, B.W., Lijk, L.J. and Hol, W.G.J. (1981). J.Appl.Cryst. 14, 274-279.
Brown, I.D. (1983). Acta Cryst. A39, 216-224.
Burova, E.M., Židkov, N.P., Zilberman, A.G., Zubenko, V.V., Nebutovskij, L.Š., Umanskij, M.M. and Ščedrin, B.M. (1977). Kristallografija. 22, 1182-1190.
Burzlaff, H., Böhme, R. and Gomm, M. (1978). Twente. 72-80.
Busetta, B., Tickle, I.J. and Blundell, T.L. (1983). J.Appl.Cryst. 16, 432-437.
Busing, W.R. (1981). WMIN, Rep. ORNL-5747. ORNL, Tenn. USA.
Busing, W.R., Martin, K. and Levy, H.A. (1963). ORGLS, Rep. ORNL-TM-305. ORNL, Tenn. USA.
Busing, W.R., Martin, K. and Levy, H.A. (1964). ORFFE, Rep. ORNL-TM-306. ORNL, Tenn. USA.
Calvert, L.D., et al. (1980). NBS Spec.Publ. 567, 513-535.

Chabot, A.A., Geddes, A.J., North, A.C.T. and Patterton, E.A. (1981). Acta Cryst. A37, C-340.
Cheetham, A.K. and Taylor, J.C. (1977). Solid State Chem. (1977). 21, 253-275.
Chigarov, A.V., Surikov, V.V. and Ivanova, L.N. (1981). Acta Cryst. A37, C-281.
Connolly, M.L. (1982). Ottawa. 273.
Connolly, M.L. (1983). Science, 221, 709-713.
O'Connor, B.H. and Bagliani, F. (1976). J.Appl.Cryst. 9, 419-423.
Cooper, M.J., Rouse, K.D. and Sakata, M. (1981). Z.Krist. 157, 101-117.
Coppens, P., Guru Row, T.N., Leung, P., Stevens, E.D., Becker, P.J. and Yang, Y.W. (1979). Acta Cryst. A35, 63-72.
Crowther, R.A. and Blow, D.M. (1967). Acta Cryst. 23, 544-548.
Coyle, R.A. (1981). Acta Cryst. A37, C-279.
Davies, G.R., Jarvis, J.A., Kilbourn, B.T., Mais, R.H.B. and Owston, P.G. (1970). J.Chem.Soc. A, Part II, 1275-1283.
Diamond, R. (1971). Acta Cryst. A27, 436-452.
Diamond, R. (1980). Bangalore. 27.01-27.15.
Diamond, R. (1983). Kyoto. 353-357.
Dodson, E.J., Isaacs, N.W. and Rollett, J.S. (1976). Acta Cryst. A32, 311-315.
O'Donnell, T.J. and Olson, A.J. (1982). Ottawa. 326-336.
von Dreele, R.B., Jorgensen, J.D. and Windsor, C.G. (1982). J.Appl.Cryst. 15, 581-589.
Edmonds, J.W. (1980). J.Appl.Cryst. 13, 191-192.
Egert, E. (1983). Acta Cryst. A39, 936-940.
Feldman, J.R., Bing, D.H., Furie, B.C. and Furie, B. (1978). Proc.Natl. Acad.Sci.USA, 75, 5409-5412.
Fiala, J. and Melichar, Z. (1984). Chem.listy. 78, 54-60.
Finger, L.W. and Prince, E. (1975). NBS Tech.Note 854, 1-128.
Frenz, B.A. (1983). SDP. Quick Ref.Quide.Euraf Nonius, Delft, The Netherlands.
Frevel, L.K. (1965). Anal.Chem. 37, 471-482.
Furey, W.Jr. (1983). Kyoto. 319-330.
Gabe, E.J. (1980). Bangalore. 23.01-23.15.
Gabe, E.J. and Lee, F.L. (1981). Acta Cryst. A37, C-339.
Giacovazzo,C. and Cascarano, G. (1981). Acta Cryst. A37, C-324.
Gilmore, C.J. (1982). Ottawa. 126-140.
Gilmore, C.J. (1984). J.Appl.Cryst. 17, 42-46.
Gilmore, C.J., Mallinson, P.R., Muir, K.W. and White, D.N.J. (1981). Acta Cryst. A37, C-340.
Goodacre,G.W. and Lee, J.D. (1971). Acta Cryst. B27, 1055-1061.
Hall, S.R. and Stewart, J.M. (1980). Bangalore. 22.01-22.21.
Hall, S.R. (1983). Kyoto. 309-318.
Harada, Y., Lifchitz, A. and Berthou, J. (1981). Acta Cryst. A37, 398-406.
Hare, T.M., Russ, J.C. and Lanzo, M.J. (1982). Adv.X-Ray Anal. 25, 237-243.
Harker, D. (1936). J.Chem.Phys. 4, 381-390.
Hecq, M. (1981). J.Appl.Cryst. 14, 60-61.
Hewat, A.W. (1973). UK Atomic Energy Authority Res.Group Rep. RRL 73/897.
Hoard, L.G. and Nordman, C.E. (1979). Acta Cryst. A35, 1010-1015.
Hornstra, J. (1970). Ottawa. 103-109.
Hoverstreydt, E.R. (1983). J.Appl.Cryst. 16, 651-653.
Huang, T.C. and Parrish, W. (1975). J.Appl.Phys.Lett. 27, 123-124.
Huang, T.C. and Parrish, W. (1982). Adv.X-Ray Anal. 25, 213-219.

Hubbard, C.R., Stalick, J.K. and Mighell, A.D. (1982). Adv.X-ray Anal. 25, 99-109.
Huber, R. (1970). Ottawa. 96-102.
Hughes, E.W. (1941). J.Amer.Chem.Soc. 63, 1731-1752.
Immirzi, A. (1980). Acta Cryst. B36, 2378-2385.
Ipatova, E.N., Ovchinnikov, V.E., Solovyeva, L.P. and Tschernov, A.N. (1974). Kristallografiya 19, 248.
Isaacs, N. (1982). Ottawa. 381-397.
Isaacs, N. (1983). Kyoto. 197-210.
Jack, A. and Levitt, M. (1978). Acta Cryst. A34, 931-935.
Jenkins, R. and Hubbard, C.R. (1979). Adv.X-Ray Anal. 22, 133-142.
Johnson, C.K. (1965). Rpt. ORNL-3794. Oak Ridge Natl.Lab., Tennessee.
Johnson, C.K. (1976). ORTEP-II, Rep. ORNL-5138. ORNL, Tennessee.
Johnson, C.K. (1980). Bangalore. 26.01-26.10.
Johnson, G.G. (1977). In Laboratory Systems and Spectroscopy, New York : M.Dekker.
Johnson, G.G. (1981). Acta Cryst. A37, C-277.
Johnson, G.G. and Vand, V. (1967). Ind.Eng.Chem. 59, 19-31.
Jones, T.A. (1982). Ottawa. 303-317.
Karle, J. (1983). Kyoto. 135-155.
Karle, J. and Hauptman, N. (1956). Acta Cryst. 9, 635-651.
Karle, I.L. and Karle, J. (1963). Acta Cryst. 16, 969-975.
Kennard, C.H.L. (1984). 13th Congress IUCr, Hamburg.
Khattak, C.P. and Cox, D.E. (1977). J.Appl.Cryst. 10, 404-411.
Kiers, C.T.and Schenk, H. (1983). Proc. 8th ECM, Liege, 263.
Kohlbeck, F. and Hörl, E.M. (1976). J.Appl.Cryst. 9, 28-33.
Kohlbeck, F. and Hörl, E.M. (1978). J.Appl.Cryst. 11, 60-61.
Konnert, J.H. and Hendrickson, W.A. (1980). Acta Cryst. A36, 344-350.
Ksenofontova, S.S., Melnikov, V.A., Mzhachikh, V.N. and Sirota, M.I. (1978). Kristallografija 23, 935-941.
Kutschabsky, L. and Reck, G. (1976). Praha, 131-137.
Langridge, R., Ferrin, T.E., Kuntz, I.D. and Connolly, B. (1981). Science 211, 661-666.
Langs, D.A. (1975). Acta Cryst. A31, 543-550.
Langs, D.A. and Hauptman, H.A. (1978). Twente. 113-118.
Lejeune, J., Michel, A. and Durant, F. (1983). Proc. 8th ECM, Liege, 287.
Lenstra, A.T.H. and Schoone, J.C. (1973). Acta Cryst. A29, 419-423.
Levinthal, C. (1966). Sci.Amer. 214, 42-52.
Lin Tian-Hut, Zhang Sai-Zhu, Chen Lin-Jun and Cai-Xin-Xing. (1983). J.Appl.Cryst. 16, 150-154.
Louër, D. and Vargas, R. (1982). J.Appl.Cryst. 15, 542-545.
Luger, P. and Fuchs, J. (1983). Proc. 8th EMC, Liege, 273.
Machin, P.A., Campbell, J.W. and Elder, M. (1981). Proc. of the Daresbury Study Weekend, Daresbury Lab., Sci. and Res.Council.
Malmros, G. and Thomas, J.O. (1977). J.Appl.Cryst. 10, 7-11.
Matthewman, J.C., Thompson, P. and Brown, P.J. (1982). J.Appl.Cryst. 15, 167-173.
Meyer, E. (1970). J.Appl.Cryst. 3, 392-395.
Meyer, E.F. et al. (1974). Fed.Proc. 33, 2402-2405.
Mortier, W.J. (1980). NBS Spec.Publ. 567, 315-324.
Mortier, W.J. and Costenoble, M.L. (1973). J.Appl.Cryst. 6, 488-490.
Motherwell, W.D.S. and Clegg, W. (1978). Univ.Chem.Lab., Cambridge, England.
Newman, W.M. and Sproul, R.J. (1979). Principles of Interactive Computer Graphies. New York: McGraw-Hill, 2nd ed.
Nichols, M.C. (1966). UCRI 70078, Lawrence Livermore Lab.
Nichols, M.C. and Johnson, Q. (1980). Adv.X-Ray Anal. 23, 273-278.

Nordman, C.E. (1980). Bangalore. 5.01-5.13.
Norrestam R. (1981). VANDER. Program for Raster Plot Represent. of Cryst. and Mol.Structures. Tech.Univ. of Denmark.
Norrestam, R. (1982). XTAPL Interactive APL Programs for Cryst.Struct. Calculations. Techn.Univ. of Denmark.
Osgood, B.C. and Snyder, R.L. (1980). NBS Spec.Publ. 567, 91.
Paszkowicz, W. and Rozbiewska, M. (1983). Proc. 8th ECM, 280.
Pawley, G.S., Mackenzie, G.A. and Dietrich, O.W. (1977). Acta Cryst. A33, 142-145.
Prince, E. (1980). NBS US Tech.Note No. 1117, 8-9.
Prince, E. (1981). J.Appl.Cryst. 14, 157-159.
Rietveld, H.M. (1967). Acta Cryst. 22, 151-152.
Rietveld, H.M. (1969). J.Appl.Cryst. 2, 65-71.
Richardson, D. (1978). Private comm.Univ.College London.
Riche, C. (1982). Proc. 7th ECM, Jerusalem, 25.
Rollett, J.S. (1970). Ottawa. 302-308.
Sakurai, T. and Kobayashi, K. (1979). Rigaku Kenkyusho Hokoku, 55, 69-77.
Schenk, H. (1980). Bangalore. 7.01-7.15.
Scheringer, C. (1963). Acta Cryst. 16, 546-550.
Schreiner, W.N., Surdukowski, C. and Jenkins, R. (1982). J.Appl.Cryst. 15, 513-523.
Schreiner, W.N., Surdukowski, C. and Jenkins, R. (1982). J.Appl.Cryst. 15, 524-530.
Sheldrick, G.M. (1981). Acta Cryst. A37, C-337.
Sheldrick, G.M. (1981). SHELXTL 81 User Manual, Rev.3, Nicolet XRD Corp., Cupertino, Calif., USA.
Sheldrick, G.M. (1982). SHELX 82. Program for Crystal Struct.Det. Univ. Of Cambridge, England.
Shirley, R. (1978). Twente. 221-234.
Shirley, R. (1980). NBS Spec.Publ. 567, 361-382.
Shirley, R. (1983). Kyoto. 1-27 (II).
Shirley, R. and Louër, D. (1978). Acta Cryst. A34, S382.
Sielecki, A.R., James, M.N.G. and Broughton, C.G. (1982). Ottawa. 409-419.
Simonov, V.I. (1982). Ottawa. 150-158.
Smith, A.R.R., Cheetham, A.K. and Skarnulis, A.J. (1979). J.Appl.Cryst. 12, 485-486.
Smith, G.S. and Kahara, E. (1975). J.Appl.Cryst. 8, 681-683.
Sparks, R. (1978). Twente. 52-63.
Spek, A.L. (1982). Ottawa. 528.
Stewart, J.M. (1978). Twente. 3-6.
Stewart, J.M. (1982). Ottawa. 497-505.
Stewart, J.M., Doherty, R.M., Munn, R.J., Hall, S.R., Alden, R., Freer, S.T. and Olthof-Hazenkamp, R. (1982). XTAL Programmers Manual.Tech. Rep. TR-873.1, Comp.Sci.Center, Univ. of Maryland, USA.
Stewart, J.M., Machin, P.A., Dickinson, C.W., Ammon, H.L., Heck, H. and Flack, H. (1976). XRAY 76.Rep. TR-446. Comp.Sci.Center, Univ. of Maryland, USA.
Sussman, J.L. (1983). Kyoto. 211-242.
Swanson, M., Rosenfield, R.E. and Meyer, E.F. (1982). J.Appl.Cryst. 15, 439-442.
Takano, T. (1977). J.Mol.Biol. 110, 569-584.
Taupin, D. (1973). J.Appl.Cryst. 6, 266-273.
Taupin, D. (1973). J.Appl.Cryst. 6, 380-385.
Tollin, P. (1970). Ottawa. 90-95.
Toray, H. and Marumo, F. (1980). Rep. of the Lab. of Eng.Makr.Tokyo Inst. of Tech. 5, Nagatsuta, Yokohama, Jap., 55-64.
Tovbis, S.B. and Schedrin, B.M. (1970). Kristallografiya 15, 1127-1134.

Visser, J.W. (1969). J.Appl.Cryst. 2, 89-95.
Waser, J. (1963). Acta Cryst. 16, 1091-1094.
Watkin, D. (1982). Ottawa. 351.
Weiss, Z., Krajíček, J., Smrčok, L. and Fiala, J. (1983). J.Appl.Cryst. 16, 493-497.
Werner, P.E. (1964). Z.Kristallogr. 120, 375-387.
Werner, P.E. (1976). J.Appl.Cryst. 9, 216-219.
Werner, P.E. (1980). NBS Spec.Publ. 567, 503-509.
Werner, P.E., Salmoné, S., Malmros, G. and Thomas, J.O. (1979). J.Appl. Cryst. 12, 107-109.
White, P.S. (1984). 13th Congress IUCr, Hamburg.
Wiles, D.B. and Young, R.A. (1981). J.Appl.Cryst. 14, 149-151.
Will, G., Frazer, B.C. and Cox, D.E. (1965). Acta Cryst. 19, 854-857.
Will, G. (1979). J.Appl.Cryst. 12, 483-485.
Will, G., Parrish, W. and Huang, T.C. (1983). J.Appl.Cryst. 16, 611-622.
Wlodawer, A. and Hendrickson, W.A. (1982). Acta Cryst. A38, 239-247.
Woolfson, M.M. (1983). Kyoto. 121-134.
Worlton, T.G., Jorgensen, J.D., Beyerlein, R.A. and Decker, D.L. (1976). Nucl.Instrum.Methods, 137, 331-337.
Wright, W.V. (1982). Ottawa. 294-302.
Wrinch, D.M. (1939). Phil.Mag. 27, 98-122.
Young, R.A. (1980). NBS Spec.Publ. 567, 143-163.
Young, R.A., Mackie , P.E. and von Dreele, R.B. (1977). J.Appl.Cryst. 10, 262-269.
Young, R.A., Prince, E. and Sparks, R.A. (1982). J.Appl.Cryst. 15, 357-359.
Yvon, K., Jeitschko, W. and Parthé, E. (1977). J.Appl.Cryst. 10, 73-74.
Zucker, U.H., Perenthaler, E., Kuhs, W.F. and Schulz, H. (1981). Acta Cryst. A37, C-332.

THE DESIGN, DEVELOPMENT AND IMPLEMENTATION OF PROGRAM SYSTEMS

By S.R.Hall

Crystallography Centre, University of Western Australia,
Nedlands 6009, Australia.

Abstract

Program systems are the predominant form of computer software used in crystallographic laboratories. Considerations that go into the organization of these systems will be discussed. Experience with the recently-released XTAL System (Stewart & Hall, 1983) will be used to illustrate the requirements of a crystallographic system.

1. Introduction

A 'system' is, for the purposes of this talk, any coordinated set of programs. For crystallographic calculations these range from small dedicated packages to large general systems. In this talk the organization of crystallographic systems will be considered in three parts: design criteria, system development and implementation procedures. Design criteria represent the 'specifications' of the system. Development is when the actual programming of these concepts takes place and implementation includes the logistics of documentation, testing, distribution and maintenance.

2. Design Criteria

In the past crystallographic systems have often evolved from software intended only for local use. Dissemination occurred either through the success of the software, or via the movement of its users. This process has worked well apart from the general problem of portability. For example the XRAY System (Stewart et al., 1972) started as local software for University of Washington in the early 60's. Although it was a reasonably simple system to adapt, it was ten years before portable versions of XRAY were available for a range of machine types.

The lack of portability in software developed for a local computer is not surprising. Crystallographers employ every type of computer, from micro to maxi, and do many different types of calculations. If a system is intended for general distribution then adaptability must be built into the system from the outset (Hall, 1976, 1984).

2.1 DISTRIBUTION LANGUAGES

There are only two languages suitable for scientific systems: FORTRAN77 and PASCAL. The pros and cons of these two languages are well-documented. To date all major crystallographic packages have been programmed in FORTRAN and there is no sign that this practice is about to change. In at least one respect this is a pity. Despite its universal use in scientific computing, FORTRAN has quite serious deficiences in its command repertoire and its adaptability to local computing environments (Wilson, 1983; Hall, 1984). For this reason the XTAL System uses the FORTRAN-derivative language RATMAC (Munn & Stewart, 1979) which is supplied on the release tape. RATMAC is translated into the local version of FORTRAN77. The arguments for and against preprocessor languages are discussed elsewhere (Kernighan & Plauger, 1976; Hall, Stewart & Munn, 1980; Hall, 1984). Suffice it to say that RATMAC provides for enhanced portability and machine-specificity over the conventional FORTRAN approach. Some drawbacks to RATMAC are the extra level of processing required and the unfamiliarity to most users. It is too early to access the implications of these factors on the long-term viability of the XTAL System. From the programmers point of view, however, the power and versatility of RATMAC makes it a far superior language to FORTRAN for portable software development.

2.2 SYSTEM GENERALITY

Design specifications must define the types of calculations to be performed on a system. Effectively such specifications determine the limits beyond which system developers need not consider. For instance, the decision that a system will do macromolecular calculations has a profound effect on the design criteria. Similarly, the specifications of software for 'precision density' calculations involve quite different constraints. Common sense dictates that these specifications are kept as general as possible, within the resources available. Any compromises are then made in favour of the intended 'market' of the system. Here are some of the principal considerations that go into system generality.

2.2.1 Symmetry generalization: The use of symmetry information is fundamental to all crystallographic calculations. It follows that the treatment of symmetry has a significant effect on the overall organization of a system. The way in which symmetry information is to be stored and applied must be decided at the outset. For example, most systems enter symmetry data once and it is transferred to all subsequent calculations via binary files. The importance of defining symmetry algorithms from the outset can be illustrated from the experience of XTAL development. Early versions of XTAL stored symmetry data both as matrices, and as equivalence parameters for each reflection. This procedure was based on the approach of the XRAY System (Stewart et al., 1976) and on the belief that the pre-calculation of symmetry equivalences speeds-up subsequent calculations. In present day multi-task computing environments, however, such gains in efficiency are largely offset by increased I/O and unpacking overheads. The pre-calculation of symmetry equivalence information also greatly increased disc storage requirements, especially for macromolecular data sets. The adoption of this form of symmetry generalization was a bad decision based on past practices and outmoded computing techniques. This was realized several years into the XTAL development and the decision to remove the reflection equivalences is estimated to have delayed XTAL release by up to 12 months. The lesson here is that careful planning of what appears to be a relatively trivial matter, such as symmetry generation, can avoid quite serious inefficiencies and delays.

2.2.2 Data file structures: As with most other branches of computing, crystallographers use two basic types of files - 'character' and 'binary'. Typical character files are the eye-readable line input file and line output (printer) file. The clarity of these files determines to a large extent how well the user communicates with the calculation. Some typical requirements for line input are user-friendly, free-format, conversational/ and/or interactive features. Similarly, line output files should be explanatory but concise enough to scan on a VDU screen.

Binary files, on the other hand, contain packed information that serves either as temporary storage during a calculation, or as archive file between calculations. The detailed planning of the binary file structure is of special importance in crystallographic systems because of the size and scope of data types. These specifications exert perhaps the largest influence on the organization of the system. Desirable properties for a binary file are ease-of-access, facility for variable data types and for efficient data transfer. These are desirable but elusive objectives. Data base management techniques are continually improving and review of current literature is recommended during the design stage.

The XTAL System employs a 'directory-driven' binary file structure which permits any data type, and a wide range of data lengths. Details of the XTAL character and binary file structures are given in the XTAL Users Manual (Stewart & Hall, 1983) and a general summary has been published by Hall, Stewart & Munn (1980) and Hall (1984).

2.2.3 Memory management: One of the major sources of inflexibility in existing systems is the use of multi-dimensioned arrays for data storage. This type of memory definition is very inflexible and is usually one of the first parts of a system that has to be modified for a new computing environment. It has other drawbacks as well. The size of multi-dimensioned arrays is determined by the maximum index likely to be encountered. This means that usually only part of these arrays are used to store data, resulting in unnecessarily large programs, higher memory charges and inefficiencies on computers that prefer contiguous memory usage (e.g. VMS and pipeline).

An alternative to the multi-dimensioned array method of data storage is the single one-dimensioned approach pioneered in the XRAY system (Stewart et al., 1972). The benefits of this approach are descibed by Stewart (1976). In simple terms, it provides flexibility for storing data of unpredictable length and promotes high-speed storage/retrieval by close-packing data. It is particulary well-suited to both VMS and pipeline computers. A one-dimensioned array also enables 'dynamic memory allocation' procedures, if supported. Memory management is also a concern in the organization program modules under different operating systems. This aspect of memory management determines the loaded structure of the system, and will be discussed further in 3.

2.2.4 Data limits: It is necessary to define as closely as possible the acceptable limits of specific data types. In crystallography it is usual to fix the maximum permitted values for |h|, |k|, and |l|. In general, once such limits are set, they are virtually impossible to alter at some later date. For instance, the decision in the XTAL System to set the maximum values of h,k,l at 512 may prove short-sighted if macromolecular structures continue to increase in size at the present rate, and intense radiation sources extend the range of measurable data. A future attempt to increase this maximum would require at least one man-year of programming effort.

The limits of other types of data also have to be considered. For instance, what are the maximum allowable number of independent data sets (e.g. multiple isomorphous replacement and anomalous dispersion data sets); multiple atomic scattering factors (e.g. precision density studies); atom species; independent scale factors; and maximum length of atom name strings? If possible, limits should not be placed on data types which are by nature open-ended. Examples of this are the number of reflection data, atomic sites or refinable parameters. Who knows how many reflections and atoms will be required in the analysis of a large virus structure! These data types can be accomodated by careful memory and file structure planning.

2.3 HARDWARE CONFIGURATIONS

Languages such as FORTRAN77 attempt to be independent of hardware considerations. It does not follow, however, that the system will be! It is essential to define at the outset the minimum hardware configuration that the system will run on. This includes the word length, the on-line memory size and the mimimum workable disk storage area. If machines with integer and real words less than 32 bits are permitted, this determines the numeric precision, word-packing and memory addressability of the system. Because of the general movement towards 32-bit hardware, developing a system for a smaller word is unwise except for control software.

3. Systems Development

Ideally the development of software should not start until the design specifications of the system are complete. In practice there is always overlap between the design and development stages. Early coding efforts are often required to prove if a particular design idea is feasible or not. Prototype software is also used to test specifications and to generate additional criteria. The real danger is, however, to let the overlap of design and development last too long! At a certain point it is imperative that the specifications of the system be frozen. This is an extremely difficult rule to enforce. Programmers are always finding 'a better way' that requires some small change to the design specifications. Unfortunately even small changes, especially in collaborative development, can badly disrupt the programming schedule. XTAL developers learnt that lesson the hard way!

> "The more innocuous a design change appears,
> the further its influence will extend."
>
> Murphy's Laws on Technology

3.1 COLLABORATIVE DEVELOPMENT

Collaborative development of a system involves more than one laboratory. It provides a broader financial base, and a sharing of computing and manpower resources. On the the other hand collaborative programming requires a much higher level of communication and organization than is needed for single-site development. It is also true that the more people and laboratories involved in deciding the design criteria, the more difficult it is to reach a consensus. Nevertheless the broader the input base, the more likely the final product will satisfy a wider range of users. The XTAL System was developed with direct collaboration of four laboratories, with input from a number of others. The design criteria were also strongly influenced by three meetings of the National Resource for Computing in Chemistry (USA). The cost of developing XTAL was borne by these laboratories and several funding agencies. It is worth noting that the initial design specifications for XTAL were published as a 'Programmers Guide' in late 1978, some 5 years before the full release of the system. This early publication of system specifications was because of the collaborative nature of XTAL. Unfortunately this promising start was somewhat undermined by subsequent enhancements to the specifications. Our experience is that late changes should be avoided at all costs.

> "The primary function of the systems designer is to make things as difficult as possible for the programmer and impossible for the implementor."
>
> Murphy's Laws on Technology

3.2 SYSTEM STRUCTURE

The modularity of crystallographic calculations permits a variety of system structures. One structure involves a series of 'standalone' programs which may be run quite independently of each other. Execution is usually sequenced via the local command language. This type of structure is popular because it is relatively simple to implement. Its principal drawback is its inability to make decisions during a chain of calculations, and that it requires use of the local command language.

An alternate system structure consolidates those routines that are common to most calculations. The 'nucleus' or 'kernel' of service routines might include line input/output, word manipulation, and memory assignment operations. The nucleus routines may also be used to control the sequence of calculations in a machine-independent way. This approach reduces the overall size of the system and ensures that these processes are always handled identically.

For a system that is developed collaboratively, the nucleus approach has additional advantages. It ensures a self-consistent approach to operations that often have a high degree of machine-specificity. Updates to these functions is also more consistently applied if they are 'program independent'. The XTAL System has adopted the nucleus approach in a way that allows each new program to be easily added to an existing software framework. This is achieved mainly through the precise definition of system protocols that each program must satisfy in order to communicate with the nucleus, and with other programs via binary files. Any programmer may add software to XTAL by adhering to the protocols detailed in the XTAL Programmers Manual (Stewart et al., 1983).

3.3 SYSTEM DOCUMENTATION

Adequate documentation is critical to the usability of a system. Crystallographic systems do not excel in this respect. Good documentation means different things to different people. To the user, it means a clear description of input parameters and commands. To the implementor, it also includes a detailed guide for implementing the system in different computing environments; well-commented code for subsequent updates or local modifications; sensible error messages and regular updates.

The layout of line input and output is a form of documentation. These are the character files already referred to in Section 2.2.2. Line input and output must be as self-descriptive as possible. A user should be able to scan the input file and understand exactly what it requests the system to do. Alphabetic codes should always be used instead of numeric flags. Sensible defaults should be assigned to all parameters. Ideally the user should only need to specify a parameter if it is different from that normally expected. Correspondingly, line output must explain exactly which parameters were used and, as concisely as possible, the important results of the calculation. Avoid the use of excessive jargon even though this is often hard to resist. For a more detailed review of the considerations that went into XTAL line input/output refer to Hall & Stewart (1980).

3.4 MACHINE INDEPENDENCE AND PORTABILITY

Machine independence and portability are usually foremost in any list of design criteria. It is often misconstrued that these are the same property. This belief arises largely from the practice of achieving portability by programming only in commonly available commands. This is the 'pidgin' FORTRAN approach used in XRAY and other systems.

It is, however, possible to make a system both portable and machine dependent. For instance, a system can be designed so that the machine-specific commands are flagged and therefore easily identified with a standard Editor. In this way versions of the system can be created for different computers through selective editing. In practice this approach can be quite difficult, especially when it comes to the maintenance of multiple versions.

In the XTAL System a more flexible variation of the 'edit' approach is used. Using the MACRO facility of the RATMAC language, it is possible to globally edit selected code. All XTAL code which is potentially machine-specific (e.g. input/output and word packing operations) is defined in a limited set of MACRO's. There are less than 100 MACRO's (out of >100,000 lines of code) that must be re-defined for different machines. Examples of MACRO applications are detailed in the XTAL Programmers Manual (Stewart et al., 1983) and summarized by Hall, Stewart & Munn (1980).

3.5 CALCULATION PRIORITIES

Early in the system organization it must be decided which, and in what order, calculations are to be included. In crystallography there is a reasonably logical order to program development based on the hierachy of calculations. The actual depth, or priority, of some calculations will depend both on the scope of the system and on the available manpower. Here is a simple chart of the typical calculations for an up-to-date crystallographic system.

1st priority	2nd priority	3rd priority
data reduction	sort/merge	profile analysis
direct methods	absorption	powder analysis
Fourier maps	vector search	slant Fourier
peak search	model builder	contouring
structure factor	error analysis	thermal analysis
SF least squares	rigid body SFLS	con/restrained SFLS
bond length/angle	atom coordination	file editor
molecular plotting	SCFS archiver	colour graphics

These calculations are listed here by their generic names and as such some may be combined in a single routine or represent more than one program. Specific macromolecular and precision density calculations are not included in this table.

4. Implementation Procedures

Both implementation and development stages of a system occur in parallel. Software is implemented and tested during the course of system development. Nevertheless, the considerations that go into implementation are quite distinct from those of software development. They include linking and loading modes, testing software, system distribution and maintenance.

4.1 LINKING AND LOADING MODES

The linking processes used on different machines require special consideration. There are three basic 'link' types in use today. They are standalone, overlay and virtual memory. The problem for the system developer is not so much to satisfy the different linking requirements, as to gain the maximum efficiency in each case. Adaptability at this level is non-trivial, and requires special care at the design stage. If this is not done it is possible, indeed probable, that a system optimized for one linking procedure will perform poorly under others. For example, programs intended for a standalone or VMS environment can become almost unusable on small machines which employ overlay linking procedures. Similarly, software structured for overlay or standalone execution may be quite inefficient on a VMS machine, due to poor data storage techniques. For this reason the implications of different linking methods need to be thought about well before general implementation of the system commences.

4.2 SYSTEM TESTING

The complete testing of a crystallographic system is virtually impossible. Nevertheless, each programmer must test as exhaustively as time permits. There are a number of standard tests available (Ahmed et al., 1972; Sheldrick, 1984), and these should form the basis for more extensive tests. The real test for new software is its application by uninitiated users. They quickly find problems with the documentation and software that the programmer, because of his knowledge of the program, often misses. Version testing should therefore be done by actual users, and at different sites. The ultimate test comes when the system is released. Further errors are certain to be found particularly during its first year of use. Good testing procedures are therefore closely allied with a regular maintenance schedule.

4.3 DISTRIBUTION AND MAINTENANCE

The importance of adequate distribution and maintenance procedures is often underestimated with scientific systems. This stems largely from the practice of distributing software free-of-charge, or at the cost of tapes, printing and postage. Because of a lack of support, users are often expected to fend for themselves. This 'caveat emptor' approach to distributed software has worked surprisingly well in the past, which is probably more of a tribute to the expertise of implementors, than to the reliability of the software. With future systems, developers may not be able to rely on such expertise. Many of today's users do not have, and do not need, the computing expertise available in crystallographic laboratories ten years ago. This means that implementation instructions should be directed at the relatively inexperienced user, and a higher proportion of system resources must go into distribution documentation and subsequent maintenance.

The task of distribution (printing, tape copying, packaging and postage) is time-consuming and tedious. If the local computer centre staff can handle this task at reasonable cost then this is a recommended approach. Distribution charges can be adjusted accordingly since most users and laboratories are prepared to pay more for a well-supported product. This can also provide a source of funds for financing continued maintenance. Here are some other ideas to improve system support:

* plan system updates to be distributed via a regular newsletter
* documentation updates should be considered as important as software updates
* in collaborative systems errors and corrections should be sent directly to the program authors for checking
* distribution and updates must be distributed from a single source to avoid duplication and proliferation of versions
* part of the distribution procedure should be a follow-up tape containing the latest version of the system
* users should be encouraged to contribute software to the system, but only after independent testing at another site.

5. Conclusions

Crystallographic software development must stay abreast of computing advances. The expectations of users are strongly influenced by what is available in other areas of computing. A good example of this is the types of software and user-friendly documentation currently available for personal computer. If the reader doubts that a revolution is underway, look at some recent PC manuals of the Apple MacIntosh, DEC PC350 or IBM PC. This is the level of presentation that future scientists will expect in their software and documentation. The low-budget, non-commercial and specialized nature of our systems are not sufficient reasons to ignore features that are quite achievable through better planning and organization. In the past crystallographers were recognised as the leaders in scientific software development. Let us ensure that this will remain the case in the future.

6. References

Ahmed, F.R., Cruickshank, D.W.J., Larson, A.C. & Stewart, J.M. (1972). Acta Cryst. A28, 365-393.

Hall, S.R. (1976) "Crystallographic Computing Techniques." Ed. F.R. Ahmed. pp. 424-432. Copenhagen:Munksgaard.

Hall, S.R. (1984) "Methods and Applications in Computational Crystallography." Eds S.R. Hall & T. Ashida, 343-352. Oxford: Oxford University Press.

Hall, S.R., Stewart, J.M. & Munn, R.J. (1980) Acta Cryst., A36, 979-989.

Kernighan, B.W. & Plauger, P.J. (1976) "Software Tools." Reading: Addison-Wesley.

Munn, R.J. & Stewart, J.M. (1979) "RATMAC: A Primer." Technical Report TR804, Computer Science Center, U. Maryland.

Sheldrick, G.M. (1984) Test data sets (personal communication)

Stewart, J.M., Kruger, G.J., Ammon, H.L., Dickinson, C. & Hall, S.R. (1972) Ed.s "The XRAY System of Crystallographic Programs." Technical Report TR192, Computer Science Center, U. Maryland.

Stewart, J.M. (1976) "Crystallographic Computing Techniques." Ed. F.R. Ahmed, pp. 433-443. Copenhagen: Munksgaard.

Stewart, J.M. (1976) Ed. "The XRAY System of Crystallographic Programs." Technical Report TR446, Computer Science Center, U. Maryland.

Stewart, J.M., Doherty, R.M., Munn, R.J., Hall, S.R., Alden, R., Freer, S.T. & Olthof-Hazenkamp, R. (1982) "XTAL Programmers Manual." Technical Report TR873.2, Computer Science Center, U. Maryland.

Stewart, J.M. & Hall, S.R. (1983) Ed.s "XTAL Users Manual." Technical Report TR1364, Computer Science Center, U. Maryland.

Wilson, K.G. (1983) CERN Courier, June, 172-177.

"THOU SHALT NOT CHANGE THE SPEC AFTER DEVELOPMENT HAS BEGUN"

INTRODUCTION TO THE LOCAL COMPUTERS

Richard Goddard
Max-Planck-Institut für Kohlenforschung
Mülheim a.d. Ruhr

Introduction

This lecture is intended to be an introduction to the computers and software that will be available for the work sessions. I will start with a brief description of the hardware, follow this by describing the software, and conclude the talk with details of how to log in, edit files and run the software. Your tutors and the lecturers will be able to give you more expert advice during the actual work sessions.

The Hardware

In all there are nine computers for your use during the work sessions. They range in size from the institute's VAX 11/780 down to the Rainbow 100 personal computer.

VAX 11/780

This is the computer most of you will be using most of the time, so in this talk I shall lay correspondingly more emphasis on it. You will find, however, that the operating systems on the various computers are very similar. This computer is connected to terminals in all the work session rooms.

VAX 11/730

The VAX 11/730 is also an institute's computer. The two personal computers, the Evans and Sutherland PS300 graphical display unit and our diffractometers are connected to it.

PDP 11/73

It is situated in ROOM 4. Prof. Burzlaff and Dr. Gomm will be demonstrating the CRYSTAN package on it.

PDP 11/23

It is situated in ROOM 6. Dr. Straver and Dr. van Meurs will be demonstrating the SDP package on it.

ECLIPSE

At the moment the Eclipse has not yet arrived. We hope it will be running by the time the work sessions start. If this is not the case we shall put a terminal in ROOM 3. The terminal will be connected to the VAX 11/780.

Evans and Sutherland PS 300

This graphical display unit is connected to the VAX 11/730.

Jim Pflugrath and Alwyn Jones will be demonstrating FRODO on it.

PDP 23 micro

The computer will be used for data reduction and analysis of the powder diffraction data collected with the STOE one-dimensional position sensitive detector. (ROOM 7)

Professional 300

Henk Schenk will be demonstrating the SIMPEL program package on it. It has a 10MB fixed disk and two floppy disk units.

Rainbow 100

The Rainbow has a 5MB fixed disk and two floppy disk units. This and the Professional are connected to the VAX 11/730.

The Software

In this section I have listed the software which will be demonstrated in the work sessions. It can be grouped into six major areas:

- Crystal Structure Solution
- Absolute Configuration Determination
- Crystallographic Program Systems
- Protein Crystallography
- Crystallographic Databases
- Powder Crystallography

Crystal Structure Solution

SHELX-84

SHELXS

Program call : SHELXS

Program for crystal structure solution.

Input files required : SHELIN.*** (prep. by EDT)
SHELHKL.*** (reflection file)

Output file : SHELLPT.*** (list file)
SHELCRD.*** (card punch - possibl input for PATSEE)

PATSEE

PATSEES

Program call : PATSEES

Fragment search by integrated Patterson and direct methods (E. Egert).

Input files required : PATIN.*** (prep. by SHELXS)

Output files : PATLPT.*** (list file)
PATCRD.*** (card punch - possibl input for SHELXS)

MULTAN

The program system of MULTAN (version 1978 and 1980) consists of four parts.
These parts must be called in the following order :

NORMAL - MULTAN - EXFFT - SEARCH

NORMAL80

Program call : NORMAL80

Computes normalised structure amplitudes from Fo's for input in MULTAN80.

Input files required :	NORMI80.DAT	(prep. by UNIV)
	FO***.***	(reflection file)
Output files :	NORMAL.DAT	(list file)
	MULTO80.DAT	(refl. input for MULTAN80)
	RCYC.DAT	(recycle instr.)
	FOUR.DAT	(fourier input for EXFFT80)
	RFL.DAT	(binary refl. file)

MULTAN80

Program call : MULTAN80

Multi - tangent refinement.

Input files required :	MULTI80.DAT	(prep. by UNIV)
	MULTO80.DAT	(prep. by NORMAL80)
	FOUR.DAT	(prep. by NORMAL80)
Output file :	MULTAN.DAT	(list file)

EXFFT80

Program call : EXFFT80

Computes an E-map from a set of normalized structure factors.

Input files required :	FFTRI80.DAT	(prep. by UNIV)
	FOUR.DAT	(prep. by MULTAN80)
Output files :	FFTRA.DAT	(list file)
	FMAP.DAT	(input for SEARCH80)

SEARCH80

Program call : SEARCH80

Finds the coordinates of the highest peaks in the E-map.

Input files required :	SEARCI80.DAT	(prep. by UNIV)
	FMAP.DAT	(prep. by SEARCH80)
Output files :	SEARCH.DAT	(list file)
	ATOM**.DAT	(atomic param. file)

DIRDIF

Program call : DIRDIF

The DIRDIF system consists of three parts :

DIDIF :	Direct methods for difference structures
ORIENT :	Rotation functions in vector space
TRADIR :	Translation functions in DIRDIF-fourier space

NORDMAN

NORDMAN

Structure solution package based on an exhaustive computer search of stored Patterson functions (C.E. Nordman et al.). Comprises the following programs:

NOR :	Calculates the Patterson function
VEC :	Calculates the intramolecular search vectors
ORV :	Calculates tree orientation search maps
MAPSIG :	Calculates the highest values in the orientaion search maps

Absolute Configuration Determination

GFMLX

Program call : GFMLX

Full matrix least-squares program with block diagonal, damping factor and enantiopol refinement options (H. Flack).

Input files required :	GFMLX.CDR	(prep. by UNIV)
	FO***.DAT	(reflection file)
Output files :	GFMLX.DAT	(list file)
	PA***.***	(new param. file - if wanted)
	FC***.***	(new calc. refl. file - if wanted)

SIMPEL

Direct methods program which will be running here on the Professional 300.

Crystallographic Program Systems

SHELXTL

The SHELXTL package will be demonstrated on the Eclipse.

XTAL

XTAL

Program call : XTAL ⟨input data file⟩ ⟨output list file⟩

The XTAL system is a set of computer codes for carrying out the calculations necessary for solution and refinement of crystal structures by diffraction techniques.

XTAL : Executes the XTAL system (XTAL.EXE). The input data file is assumed to have the extension .DAT The default output file is 'input'.LST. XTAL binary files A through H are assigned to files named XFILE.A through XFILE.H. The ASCII punch file is assigned PUNCH.DAT .

DX : Displays the directory of files XFILE.A through XFILE.H and the PUNCH.DAT file.

PURXTL : Deletes all XFILE and PUNCH files.

CRYSTALS

Program call : CRYSTALS or STARTCRYS

Crystal structure package (interactive, on-line or batch mode).

NRC VAX Crystal Structure System

NRC

System for solution and refinement of crystal structures. (E.J.Gabe et al.)

Following programs belong to the system:

ABSORP	CDEDIT	CDFILE
CREDUC	DATRD2	DISPOW
ERRANL	FOURR	LSTSQ
MULTAN	PACKER	PLTMOL
TABLES	UNIMOL	UTILITY

After you have used the command 'NRC' you can start the programs by typing their names.

CRYSTAN

CRYSTAN is a crystallographic program system for the PDP 11/73 computer.

SDP

Structure Determination Package running both on the PDP 11/23 and the VAX 11/730.

Protein Crystallography

EREF

Program for the simultaneous refinement of intensity data and energy information.

PROLSQ

Protein Least-Squares Refinement Program.

CORELS

CORELS is a computer program primarily intended for macromolecular model building or X-ray refinement especially when one is interested in holding parts of a structure in a constrained conformation while allowing flexibility at the joints (Sussman et al. 1977).

FRODO

Runs on the Evans and Sutherland PS300

Crystallographic Databases

The Cambridge Database

CAMBRIDGE

Program call : CAMBRIDGE

The Cambridge system comprises three programs for retrieval of information from the Cambridge database and two for calculations using the atomic coordinates. (Crystal structures containing at least one organic carbon atom.)

Input prepared with CAMBRIN.

BIBSER

Bibliographic search of the file SBIB to yield :
- listings of bibliographic entries
- sub-files of bibliographic entries
- sub-files of reference codes

Input file required :	INBIB.***	(prep. by CAMBRIN)
Output files :	BIB.***	(list file)
	SUBBIB.***	(binary, used as sub-file of SBIB)
	REFCO.***	(reference codes)

CONSER

Connectivity search of the file SCON to yield :
- listings of reference codes
- sub-files of connectivity entries
- sub-files of reference codes

Input file required :	INCON.***	(prep. by CAMBRIN)
Output files :	CON.***	(list file)
	SUBCON.***	(binary, used as sub-file of SCON)
	REFCO.***	(reference codes)

RETRIEVE

Program which reads a sub-file of reference codes and retrieves the corresponding entries from SBIB, SCON, or RDAT file.

Input files required :	REFCO.***	(reference codes prep. by BIBSER or CONNSER)
Output file :	SUB.***	(input for GEOM / PLUTO)
	SUBBIB.***	(input for BIBSER - if wanted)

SUBCON.*** (input for CONNSER - if wanted)

PLUTO Program call : CAMBRIDGE

Plot program from Cambridge crystallographic set of programs. Plots molecular and crystal packing diagrams in various styles.

Input files required : PLOTIN.*** (prep. by CAMBRIN or EDT)
SUB.*** (output of RETRIEVE, or free format data)

Output file : PLOT.*** (list file)

GEOM Program call : CAMBRIDGE

The principal operations of GEOM are:

- calculation of intramolecular geometry
- calculation of intermolecular geometry
- calculation of coordination sphere geometry
- calculation of centroids, vectors, planes
- definition of a chemical fragment
- automatic or specified tabulation of geometrical parameters of the chemical fragment
- output of atomic coordinates for use with other programs
- output of atomic coordinates as line printer plot positions

Input files required : INGEO.*** (prep. by CAMBRIN or EDT)
SUB.*** (output of RETRIEVE, or free format data)

Output files : GEO.*** (list file)
ATCO.*** (atomic coordinates)

The Inorganic Database

ICSD

Program call : CRYSTIN

The ICSD system is an on-line interactive program for retrieving information from the Inorganic Crystal Structure Database prepared by the Institute of Inorganic Chemistry of the University of Bonn. It also retrieves information from the Cambridge organic and organometallic crystal structure database.

The Crystal Structure Search and Retrieval

CSSR

Program call : CSSR

The CSSR program is an on-line interactive program for retrieving information from the database of organic and organometallic crystal structures prepared by the Cambridge Crystallographic Data Centre.

Utilities

COPY

DCL (Digital Command Language) command.

COPY input-filename output-filename

RENAME

DCL (Digital Command Language) command.

RENAME input-filename output-filename

UNIV1

Program call : UNIV

XRAY utility program; prepares the input for the following programs:

DAESD	EXFFT	(G)FMLS	GBDL
GFMLX	MULTAN	NORMAL	ORTEP
PLANE	SEARCH	XANADU	

Input file required : PA***.*** (atomic param. file)

Output file : FFTRI80.DAT or
GFMLX.CDR or
MULTI80.DAT or
NORMI80.DAT or
SEARCI80.DAT or

UNIV2

Program call : UNIV2

XRAY utility program; prepares the input for the following program systems :

CRYSTALS - writes an input file for CRYSTALS using a PA file or dialogue,

MPI - reads a REFL.FOA file (SDP format) and writes a FO file (MPI format),
- reads a STRUC.NME and PARA.DTA file (SDP format and writes a PA file (MPI format),
- reads a standard crystallographic file (SCF) and writes a PA- and FO file (MPI format),

SCF - reads a PA file and FO file (MPI format) and writes a standard crystallographic file (SCF),

SHELX - reads a PA file (MPI format) and writes an input file for SHELX.

Logging in

To log into the VAX11/780 one simply types the program name: This can be one of the following:

EREF	PROLSQ	CORELS	NORDMAN
SHELX	PATSEE	MULTAN	ENANTIO
CRYSTALS	NRC	XTAL	
CAMBRIDGE	CSSR	ICSD	DIRDIF

E.g. Username: EREF
Password: SHOP

If you get the 'Request:' message, type: TS

To get onto your subdirectory, type G followed by the number you have been assigned:

e.g. G4

To return to the main directory type:

MAIN

Note: not all program systems require individual work areas, so check with your lecturer or tutor before working on a particular area.

The Editors

EDT

This is probably the editor which will be most popular. If you wish to edit or create a file you need to type:

EDI filename.extension

e.g. EDI TEST.DAT

You will then automatically be in keypad editing mode. This means that you will be able to edit the file using the both the main tabulator and the keypad on the right-hand-side of the keyboard. The HELP key (PF2) will give you advice if you have problems.

Some simple commmands are as follows:

(a)Insertion: type characters to be inserted
(b)Deletion: the DEL key will delete characters in front of the cursor

The cursor can be moved using the arrow functions.

To exit from the editor and write the new file on the disk, type the GOLD key (PF1), the COMMAND key (7) and to the question ´Command:´, type EX followed by the ENTER key.

If you prefer to work in line mode, type (CTRL)Z and then:

HELP for further information.

C returns you to keypad mode.

TECO

The TECO editor is available for those who are familiar with it. To edit a file with TECO, type:

TECO filename.extension

Logging out

We recommend that the participants use the command BYE since this effects a selected purge before the job is logged off. If this is not desired, make sure you have deleted as many files from your disk space as possible and then type LO.

Problem Datasets

There will be a selection of problem data sets available. Details about the individual problems can be obtained from Klaus Angermund. Once you have chosen a problem dataset you can copy the observed structure factor file to the disk area you are working on with the COPY command:

```
COPY
From: [XRAY.PROBLEM]*.NOZ
To: *
```

where Z is the problem dataset number.

Disk Quotas

Each user (i.e. program package) has been allocated 15000 blocks (7.7 MB) of disk space on the VAX 11/780. (One VAX block equals 512 bytes.)

If a particular area becomes full, the following error message will be output:

Disk Quota Exceeded!

It will not be possible to run any further jobs from that area until room has been made on the disk:

DEL filename.extension;version number

or PURGE filename.extension
which deletes all but the final version of a file.

Finally, in case of any problems, type:

HELP

and this will give you details of all the systems installed at Mülheim.

THE N.R.C. VAX CRYSTAL STRUCTURE SYSTEM

E. J. Gabe, F. L. Lee and Y. Le Page
Chemistry Division
National Research Council of Canada
Ottawa, Canada K1A 0R6

The routines contained in this system comprise an integrated set of programs to perform all computations necessary for the solution and refinement of crystal structures from X-ray diffraction data.

The system has been written entirely in Fortran to run in an interactive fashion. All routines are menu driven with simple input which can usually be defaulted. The system is completely symmetry general, treating symmetry constraints automatically. It is set up to run with up to 300 atoms and an essentially unlimited number of reflections. It contains routines for plotting based on Tektronix equipment and the PLOT10 library.

All code has been processed by the routine 'PRETTY' and is therefore standard Fortran, DO-loops are indented and statement labels increase in a fixed monotonic sequence.

System Data Files

The system is based on two direct access files. Direct access was chosen because of convenience in manipulation and speed of input/output.

The two files (Fig. 1) are usually referred to as the

Crystal Data file (.CD), and the

REflection file (.RE).

The .CD file contains information about the space group, cell parameters, symmetry, scattering factors and general parameters. The .RE file, as its name implies, contains reflection specific information.

Both files use 32-bit variables exclusively. The .CD file contains 8 general records, n atom records and 1 trailing record. Each record is 100 variables long. The RE file contains m records for reflections plus 1 trailing record. Each record is 32 variables long, and if necessary information about h,k,l and -h,-k,-l reflections is stored in the same record.

Several routines use scratch files and write line-printer output files, which can be safely discarded after examining. Two files which should be kept - at least temporarily - are the peak file (PEAK.DA) from the Fourier routine and the phase file (MULFOU.DA) from MULTAN. These can of course be deleted when they have no further use.

System Structure and Routines

The interconnections between system routines and files are shown in Fig. 1.

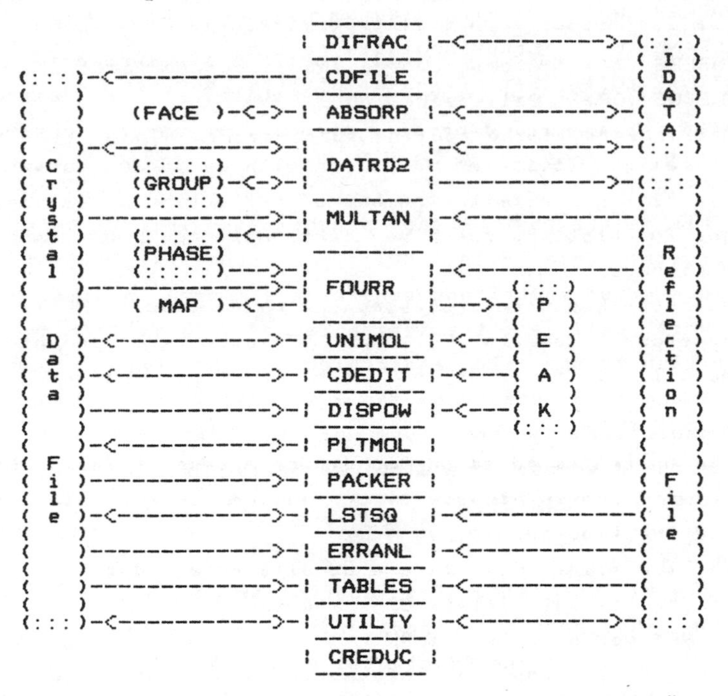

Fig. 1. Interaction of the routines with the system files. The routines are outlined as ¦ NAME ¦ and the files as (Name). The direction of data flow is shown by < and >.

The 15 main routines are well integrated and communicate for the most part via the .CD and .RE files. The routine DIFRAC is the NRC diffractometer control program, which is not distributed with the package. The routines MLTN80 and QTAN are runnable (at NRC) but are not as well integrated as the main routines. These routines will eventually be incorporated, but in the meantime MULTAN itself is available in an enhanced version which has proved to be very successful.

The 15 main routines are listed in the table below, which is followed by a short description of each.

Main Routines

1.	CDFILE (CD)	Generate the initial .CD file
2.	ABSORP (AB)	Gaussian absorption correction
3.	DATRD2 (DA)	Intensity data reduction
4.	MULTAN (MU)	Multan - local version
5.	FOURR (FO)	Fourier calculation
6.	UNIMOL (UN)	Unique molecule builder
7.	CDEDIT (ED)	Edit the .CD file
8.	DISPOW (DS)	Distance, angle and powder pattern
9.	PLTMOL (PL)	Molecular plotting routine
10.	PACKER (PA)	Packing diagram routine
11.	LSTSQ (LS)	Structure factors and least squares
12.	ERRANL (ER)	Error analysis routine
13.	TABLES (TA)	Atomic coords and structure factor tables
14.	UTILTY (UT)	Set of system utility routines
15.	CREDUC (CR)	Cell reduction

(XX) refers to the routine subdirectory name.

1. CDFILE

This is the routine that produces the initial version of the crystal data file (.CD) for a structure. It interprets the space group symbol and generates symmetry information from it. Unit cell and scattering factor data is also processed and written to the file header blocks. The space group code is an upgraded version of the SPGP code from the LASL system which processes the short form of the international symbol with each discrete operator separated by a blank, e.g. P 21 21 21 for P212121.

2. ABSORP

This routine calculates absorption corrections by the Gaussian integration method. It can accomodate up to 8000 integration points and 40 crystal faces. The input file is the IDATA file from the diffractometer program and the crystal faces are defined in terms of the angles Chi and Phi to put a face in the position opposite to the diffracting position, i.e. -Chi and 180+Phi, from the diffracting position, and d the perpendicular distance from the centre of rotation to the face in cms. It is possible to plot the crystal shape for checking and then edit the input if wanted. Trial calculations can be performed before starting the main run. The results are written to the IDATA file and applied by DATRD2.

3. DATRD2

The data reduction routine accepts intensity data from the NRC diffractometer routine, a standard Picker FACS-1 or a Nonius CAD-4 to make a reflection file (.RE). It can process multiple data sets and distinguishes between F(hkl) and F(-h-k-l) where necessary. Reflections may be scaled in several ways and weights are calculated from counting statistics or internal consistency for multiple data sets. Statistics are accumulated for internal agreement between symmetry related measurements and for the K(s) curve, from which E and epsilon values are derived and written to the .RE file.

4. MULTAN

This is a somewhat modified early version of the routine of the same name. It accepts input from the .CD and .RE files and can accomodate up to 500 E-values with up to 84 Sigma2 relationships per reflection. The original routine used scratch files quite extensively and in consequence was fairly slow, but very successful. It has recently been modified to remove the scratch files and is now much faster. The phase sets are written to a file (usually MULFOU.DA) for input to the Fourier routine.

5. FOURR

FOURR calculates all types of Fourier maps. It has many options which may be selected from the terminal. These include origin removal and sharpening for Patterson maps; Sim weighting for Fourier maps; correction for anomalous dispersion effects; a variety of print suppression and format control options if the map is to be printed; a peak picking section with 27-point fitting which eliminates repeated peaks. Almost all data manipulation is performed in memory and the routine is quite fast.

6. UNIMOL

UNIMOL takes positional data from a .CD file and/or a PEAK file and makes a reasonable attempt to make a connected unique molecule. Interatomic distances are calculated, printed and stored in the .CD file for use in PLTMOL.

7. CDEDIT

As its name implies this routine allows the user to edit the direct access .CD file. It is highly interactive and menu-driven. Atomic information can be entered manually or read from a PEAK file from the Fourier routine. Symmetry constraints are generated for input atoms and stored in the .CD file. It is the user's responsibility to fix the appropriate parameters in polar space groups.

8. DISPOW

This routine is an amalgam of a distance and angle routine and a comprehensive powder pattern calculator. Distances can be computed with or without angles and e.s.ds. Distance and angle limits may be set on individual atoms though this is not normally necessary. Powder patterns may be calculated with d-spacings and intensities in several different standard forms.

9. PLTMOL

This routine uses data from a .CD file to draw stereo pictures of organic molecules on Tektronix 4000 series terminals. The drawings can also be transferred to a 4663 plotter. It is extremely fast and has proved to be very useful in the solution of structures and in the interactive preparation of drawings for publication. Atoms in the file may be deleted and a new .CD file written. UNIMOL must be used in order to make a unique molecule the first time PLTMOL is used. Once the molecule is assembled however PLTMOL can generate its own connections.

10. PACKER

This routine prepares packing diagrams using the same devices as PLTMOL. The routine can deal with all types of molecular symmetry (I think!) and is useful for illustrating packing details and H-bond schemes. Unit cell box limits can be set and molecules are included in the plot if any atom falls within the limits of the specified unit cell box. Pictures can be further editted by the deletion of specific molecules and specific bonds may be inserted to emphasize particular points.

11. LSTSQ

This is the general structure factor and least squares calculation routine. The routine will allow refinement of up to 600 parameters by the full matrix method or 1000 parameters in the block diagonal mode. Subject to these constraints up to 300 atoms at a time may be included. In the blocked mode, atoms are not linked and up to 10 parameters/atom may be refined - occupancy, x, y, z and 6 uijs. The precise path of the calculation, number of refinement cycles and type of output is controlled from the initial dialogue. The progress of refinement can be checked by a judicious use of the printout control parameters for 'bad' reflections. Many sequences of refinement cycles may be undertaken in the same program run. For larger refinements, with the full matrix method, the cpu time and elapsed time required may become very lengthy and the routine may be run as a batch job. There is also a version for the Analogic AP500 array processor, which is approximately 10 times faster than the VAX 11/780 version with floating point accelerator.

12. ERRANL

This routine calculates weighting and agreement analyses as functions of Fo, Fo**2, sin theta/lambda, h, k and l. It should be possible to use this analysis, in conjunction with the kFo**2 weighting scheme option in LSTSQ, to obtain a uniform distribution of residuals.

13. TABLES

Tables of parameter values and structure factor listings for publication are prepared by this routine. Two types of parameter table are allowed, a) where the routine suggests the number of decimal places to use which the user can modify, and b) where the routine prints the table according to the esds in the .CD file in order to preserve reasonable significance - this is the preferred option.

14. UTILTY

This is a collection of general purpose routines for the manipulation of direct access files and some other small routines which have been found useful. It is our current practice to add small useful routines to this package in order that they may become generally available.

At present the following routines are available:

a) REMOD - Modify the contents of an .RE file.

b) REFLST - List specified reflections in an .RE file.

c) CDDUMP - Dump the contents of a .CD file to the terminal, labelling each item.

d) CDREFM - Convert .CD and/or .RE direct access files to ASCII format, to facilitate transfer between computers.

e) CDREBN - Convert ASCII .CD and/or .RE files back to direct access files.

f) FLCOPY - Copy specified blocks of one direct access file to another. This is convenient for constructing .CD files from fragments of 2 or 3 other .CD files.

g) SHORTN - Shorten an .RE file to save processing time. There are several options for deletion of reflections.

h) SCATFS - Make up composite scattering factors from 2 normal scattering factors, and insert the values into a specified place in an .RE file.

i) CDMOD - Modify the contents of a .CD file.

j) DIFFM - Convert an IDATA file to ASCII format for transmission between computers.

k) DIFBN - Convert an ASCII IDATA file back to direct access form.

l) NUNAME - Rename atoms in a .CD file. This is a much more convenient way of renaming many atoms than the method in CDEDIT.

m) EQUIVS - Compute short contacts between specified atoms or types of atoms and optionally move groups of atoms with the equivalent position transformations detected.

n) TORSHN - Calculate all torsion angles for the atoms in a .CD file.

o) LITLFM - Convert an .RE file to a compressed ASCII form for storage.

p) LITLBN - Convert a compressed ASCII .RE file back to binary.

q) NONIUS - Convert a Nonius CAD-4 intensity data file to an IDATA form for processing by DATRD2.

r) EFLMOD - Examine and/or modify specified variables in any direct access file.

s) BESPLN - Calculate least squares best planes through sets of atoms by Ito's method. Sigmas are calculated.

15. CREDUC

This routine performs a cell reduction on typed input data based on its ability to detect possible 2-fold axes in directions with low indices (Le Page, Y. (1982). J. Appl. Cryst. 15, 255.). It has successfully processed hundreds of examples - including all standard tests - and is believed to be error free.

A typical initial sequence for a structure might be as follows

1. Prepare NAMES.ZZ. (File of .CD and .RE filenames)
2. Run CDFILE.
3. Run DATRD2 (all stages).
4. Run MULTAN.
5. Run FOURR to get PEAK.DA.
6. Run UNIMOL on .CD and PEAK.DA.
7. Run PLTMOL and write out a reasonably cleaned up molecule.
8. Run LSTSQ for structure factors or least squares.
9. Run FOURR.
10. Run CDEDIT to put in a few more peaks.
11. Recycle to 6.

This is by no means a hard and fast sequence and it might be useful to try out a few routines, eg. CDFILE, CDEDIT and UNIMOL to become more acquainted, before beginning any real structure work.

<u>Timing</u>

Some typical times for different routines are as follows. Space-group P-1, 62 atoms, 5200 reflections, 460 parameters.

VAX 11/780 with Floating Point Accelerator.

a).	MULTAN	480 E-values, 37000 Sigma2 relationships.	
		Parts 1 and 2	6 mins
		Part 3	15 secs per solution.
b).	FOURR	100,000 points (Including peak search)	2 mins
c).	LSTSQ	Full matrix method VAX	60 mins per cycle
		Full matrix method AP500	8 mins per cycle
		Blocked matrix method	12 mins per cycle

CRYSTALLOGRAPHIC ALGORITHMS FOR MINI AND MAXI COMPUTERS

By George M. Sheldrick,

Institut für Anorganische Chemie der Universität, Tammannstraße 4,

D-3400 Göttingen, Federal Republic of Germany

Introduction

Computers are frequently divided into categories such as "micro", "mini", "supermini" etc. A few years ago, micros had a word length of 4 or 8 bits, minis 16 bits, and maxi (or "mainframe") computers 32 bits or more. Since the amount of memory which could be addressed efficiently and the speed of arithmetic were both dependent on the word length, it was relatively easy to assign a computer to a paticular category. With the advent of micros with a 16-bit data bus and 20 (or more) bit address range, the situation has become less clear, so it is more convenient to measure performance in the new S.I. unit, the "VAX". 1 VAX corresponds to one VAX/780, with floating point accelerator, basing the comparison on typical crystallographic operations such as least-squares refinement. For small-molecule structure determination, an allocation of about 0.1 VAX per crystallographer is sufficient. Most micros are still one or two orders of magnitude slower than this (though they are becoming faster), and the 12-bit PDP-8 or 16-bit LSI 11/23 (both about 0.02 VAX) are also scarcely adequate. However the fastest 16-bit minis (DG Eclipse S-140 0.25 VAX, Eclipse S-280 ca. 0.4 VAX) are faster than the slower 32-bit VAX'es (the VAX 730 with FPU rates at 0.20 VAX), and the fastest "superminis" (e.g. the Norsk Data 570 series at about 2 to 4 VAX) are the equal of many university computer centres (e.g. IBM 370/158 ca. 2 VAX, UNIVAC 1110 ca. 3 VAX).

In this talk I shall concentrate on algorithms designed to make the most efficient use of both mini and maxi hardware, and on the two potentially slowest stages of crystal structure determination, namely the solution of the phase problem, and least-squares structure refinement.

Software and Hardware Array Processing

In the multisolution approach to direct methods, made popular by the widely used program MULTAN, the phases of the reflections with largest E-magnitudes are refined (usually by the tangent formula) using a set of phase relations. This is repeated for a large number of starting sets, and the best solution(s) selected according to one or more figures of merit. In MULTAN only the triple phase relations are used actively in this way; MITHRIL (Gilmore, 1984) and SHELX-84 also allow negative quartets to be used for refinement purposes. A major problem arises in implementing the multisolution method efficiently on a 16-bit mini-computer: in practice the available memory is only sufficient to hold a modest number of reflections and phase relations, and disk transfers are relatively slow. In the SHELXTL mini-computer system, two alternative approaches are available, which use different algorithms to overcome the problem:

1. In the SOLV procedure, preliminary refinement of not more than 200 phases with not more than 2000 phase relations is followed by the calculation of "early figures of merit"; a small number of "best" solutions are then developed further by the (much slower) procedure of reading phase relations from the disk. This method is computationally very efficient for small and medium sized structures, but is less successful than RANT for very large structures.

2. In the RANT procedure, the phase relations are stored on the disk, but are used to process a number of phase permutations in parallel. Thus the time penalty for disk transfers, unpacking phase relations, etc. is divided by the number of parallel permutations. With this method there is effectively no limit to the number of phase relations which may be used, which makes it more suitable for large structures, although it tends to be slower than SOLV.

This technique of reducing overheads by processing several phase sets in parallel, which might be called "software array processing", saves time even when there is enough memory to store the phase relations as well, and so was incorporated into the SHELX-84 program for mainframe computers. In SHELX-84 the triple phase relations and negative quartets are first arranged into a "phase development pathway" for optimum efficiency of the phase refinement. Although the phase development pathway is stored on the disk, on a virtual memory machine it could with advantage reside in a large "virtual" array, since it is accessed sequentially and as little as possible. The critical FORTRAN code is given below; to save space only the calculation involving triple phase relations is given, the negative quartet part being entirely analogous. There are separate critical loops for centrosymmetric and non-centrosymmetric structures. NQ is the number of phase sets handled in parallel (typically 64).

```
C
C ** CRITICAL CENTROSYMMETRIC LOOP **
C
      DO 41 K=1,NQ
      D(K)=D(K)+A(I)*A(J)
      I=I+2
   41 J=J+2
```

The phase relations are decoded before the critical loops. In the centrosymmetric case adjacent elements of A store $E_i s_i$ and $-E_i s_i$, where s_i is the sign of reflection i. These values for one reflection for each of the NQ permutations thus occupy 2NQ consecutive elements of A; this storage method is advisable in case A goes "virtual", e.g. on a VAX. Decoding a phase relation thus involves setting I to the index of the first element of A which stores reflection i. Similarly J points to $E_j s_j$ for the first phase permutation; if there is a 180° phase shift in the phase relation, then 1 is added to J so that it points to $-E_j s_j$. This saves one multiplication inside the critical loop ! The sigma-2 sums sigma$[s_i s_j g E_i E_j]$ for each permutation (where g is +1 for zero phase shift and -1 for a 180° phase shift) are accumulated in array D.

```
C
C ** CRITICAL NON-CENTROSYMMETRIC LOOP **
C
      DO 43 K=1,NQ
      Y=R*A(I+1)
      X=U*A(I)-V*Y
      Y=U*Y+V*A(I)
      Q=S*A(J+1)
      D(K)=D(K)+X*A(J)-Y*Q
      E(K)=E(K)+X*Q+Y*A(J)
      I=I+2
   43 J=J+2
```

In the non-centrosymmetric case, E.cos(phi) and E.sin(phi) are stored in consecutive elements of A, so again one reflection for all permutations occupies 2NQ consecutive elements of A. The phase relation is decoded to set R to the sign which multiplies phi(i), S to the sign which multiplies phi(j), U to the cosine of the phase shift t and V to its sine. Thus D accumulates $E_iE_j$cos(r.phi(i)+s.phi(j)+t) and array E sums $E_iE_j$sin(r.phi(i)+s.phi(j)+t), as required for the tangent formula. As a weighted tangent refinement is performed, all the stored E values are multiplied by weights between 0 and 1, but the above computer code is not affected by this. Multiple dimension arrays and complex arithmetic are _never_ used in SHELX-84.

Such "software array processing" provides a good basis for programming a hardware array processor. In fact the CFT compiler on the CRAY-1 automatically vectorises the above code, improving the phase refinement performance from 35 VAX (not vectorised) to 136 VAX (vectorised).

Patterson Interpretation

Although about 30% of all crystal structures are solved by the "heavy atom method", the heavy atoms usually being found by interpretation of the Patterson function, there appear to be no widely distributed programs which attempt to automate this method. Many of the difficulties of automation may be avoided by _not_ storing the calculated Patterson map ! The peak list contains a great deal of useful information, provided that we remember that peaks may overlap, and that weaker vectors may be missing. As a final check, the Patterson function can be _recomputed_ at selected points _X_. The following algorithm, incorporated in somewhat different forms in the mainframe program SHELX-84 and the mini-computer version SHELXTL, is general for all space groups and does not require any structural information in advance. However it assumes that for _at least one_ heavy atom, _all_ Harker peaks are present in the peak list.

1. A sharpened Patterson with coefficients $[E^3F]^{1/2}$ is calculated on a grid determined by the resolution of the data, and a sorted peaklist is generated by parabolic interpolation.

2. The peak list is analysed to generate sets of atomic coordinates $\underline{x}$ for which all Harker vectors (i.e. vectors between an atom and all its symmetry equivalents) are present, to within a tolerance determined by the resolution of the data. In space group $P\bar{1}$, every peak is considered to be a potential $2\underline{x}$ vector. In P1, just one "atom" is generated, at the origin.

3. The peak list is used again to find a figure of merit for each potential atom $\underline{x}$ generated by stage 2. First a list of coordinates $\underline{x}'$ is found for which every cross-vector with $\underline{x}$ is present in the peak list; all cross-vectors and Harker vectors involving pairs of "atoms" $\underline{x}'$ are then calculated, and compared with the peak list to estimate the figure of merit.

4. The potential atom $\underline{x}$ with the best figure of merit is used as a basis for stage 6. If required, one or more extra atoms are added from the $\underline{x}'$ list used in stage 3. This is essential in space group P1, and may be desirable in a few other low symmetry cases. Alternatively the user may input one or more trial atoms $\underline{x}$ himself for use in stages 5 and 6, in which case stages 2, 3 and 4 are bypassed.

5. The heavy atom coordinates may be refined by moving each atom in turn to optimise the agreement between all vectors in which it is involved and the peak list. In practice this is only useful when a set of several heavy atoms was generated or input in stage 4.

6. All peaks $\underline{X}$ are combined with the heavy atom(s) and their symmetry equivalents $\underline{x}^*$ to generate further possible atomic positions $\underline{x}'' = \underline{x}^* \mp \underline{X}$. The list of $\underline{x}''$ which arise most frequently is optimised and used with the heavy atom(s) $\underline{x}$ to generate a "crossword table":

```
x1   y1   z1     M11
                 d11

x2   y2   z2     M22   M12
                 d22   d12

x3   y3   z3     M33   M13   M23      etc.
                 d33   d13   d23
```

where M_{ij} is the minimum sharpened Patterson density $P(\underline{x}_i-\underline{x}_j^*)$ and d_{ij} is the minimum distance $|\underline{x}_i-\underline{x}_j^*|$. P is recalculated from the reflection data, not derived from the peak list. In pseudosymmetry problems (and also where the enantiomorph has not yet been defined) the atom list contains double or multiple images of the structure, which can be resolved by finding a set of atoms linked by consistently high M_{ij} values. Only in this final interpretation is use made of chemical information (i.e. atomic numbers and distances), so unexpected structures are also easy to recognise. The generation of the crossword table is the slowest stage in the procedure, but vectorises automatically (from 33 VAX to 150 VAX) on the Cray-1.

Example: the final table for a structure "pyAuBr" in $P\bar{1}$ was:

x	y	z	self	cross-vectors			
.209	.488	.097	246.5 3.52	Au(1)			
.286	.787	.278	268.1 5.87	457.2 3.25	Au(2)		
.991	.168	.835	19.1 4.41	156.0 3.83	141.1 2.40	Br(1)	
.560	.746	.395	19.5 4.79	149.6 3.95	140.3 2.36	49.5 4.44	Br(2)

The structure is clearly $[Aupy_2]^+[AuBr_2]^-$ with Au-Br about 2.38 Å. Note that the minimum Br(1)...Br(2) distance of 4.44 Å is intermolecular - the anion is in fact linear.

This method has been tested on about 40 known structures in a wide variety of space groups (including 5 protein isomorphous delta-F datasets), and at least as many unknown structures. Only one of the known structures failed completely (APAPA: 2 P and 67 C, N and O in the asymmetric unit in $P4_32_12$), but in two other cases the "second best" heavy atom proved to be correct, and had to be input by hand to generate the correct crossword table. One of the proteins could only be solved in P1, because one Harker peak was rather weak, and there was only one heavy atom in the asymmetric unit (with low occupancy).

Efficiency in Least-Squares Refinement

The time taken by any least-squares refinement algorithm rises steeply with the number of parameters. This number may be reduced to a minimum by means of chemically sensible constraints. For example, it is usually possible to calculate hydrogen positions more accurately from geometrical considerations (possibly incorporating simple force-field ideas) than is possible by refining them using X-ray data alone. A particularly convenient constraint for CH and CH_2 groups is the riding model, defined by the equation:

$$\underline{x}_H = \underline{x}_C + \text{constant vector}$$

which makes the calculation of derivatives for least-squares refinement particularly simple. Although only the C-H distances and H-C-H angles remain rigorously constant, in practise the procedure of alternating coordinate idealisation with riding refinement converges rapidly. If the isotropic temperature factor of each hydrogen is fixed at (say) 1.2 times the equivalent isotropic U of the carbon to which it is attached, then no least-squares parameters are needed for these hydrogen atoms. Methyl groups may be refined as rigid groups, at the cost of one (torsion) or three (torsion and tilt) extra parameters. Rigid groups may also be employed for non-hydrogen atoms; a particularly elegant and efficient method of fitting a general rigid group (from e.g. literature data or force-field calculations) to e.g. Fourier peaks has been described by Mackay (1984). In this quaternion method, the translation is found as usual by superimposing centroids; the necessary reorientation is then described as a single rotation about a general axis direction, which can be found in one step (together with the angle) by linear least-squares.

The available reflection data (and possibly also computer resources) may be inadequate to allow full anisotropic refinement of large organic or organometallic structures. On the other hand some allowance for anisotropic motion (or possibly small local disorder induced by partial solvent occupancy) may still be desirable, especially for peripheral atoms. One computationally convenient compromise is to refine axially symmetric thermal ellipsoids, with components $\bar{U} + (2\Delta U/3)$ along the axis and $\bar{U} - (\Delta U/3)$ perpendicular to it. The axis direction may be fixed by the bonds to other elements (excluding H); if there is one bond, it defines the axis, and if there are two, the axis is taken as the direction perpendicular to both. Zero or more than two bonds revert to isotropic. This provides a first approximation to typical thermal motion behaviour, at the cost of only two thermal parameters ($\bar{U}$ and ΔU) per atom. Pure restraints, such as forcing the U components along bonds to be approximately equal, are chemically easier to justify, but much more expensive, since all 6 U_{ij} parameters must be refined. However a similar three-parameter scheme, including restraints, has been described by Konnert and Hendrickson (1980).

It is instructive to look closely at the inner loop of a least-squares program, the addition of derivative products to the matrix. Computer time and memory can be saved by storing the matrix in lower triangular form in the one-dimensional array B, and the weighted derivatives $w_i^{1/2}(\partial F_i/\partial p_j)$ for a given reflection i in array D. The following coding then appears at first sight to be optimal (n.b. it is advisable to store the number of parameters N in a local rather than a COMMON variable):

```
      M=1
        DO 99 J=1,N
        T=D(J)
          DO 99 K=1,J
          B(M)=B(M)+T*D(K)
99        M=M+1
```

In fact, this algorithm wastes about half its time incrementing and testing the loop variables. The program can be made more efficient by adding to the matrix only every (say) 5 reflections:

```
      M=1
        DO 99 J=1,N
        T1=D1(J)
        T2=D2(J)
        T3=D3(J)
        T4=D4(J)
        T5=D5(J)
          DO 99 K=1,J
          B(M)=B(M)+T1*D1(K)+T2*D2(K)+T3*D3(K)+T4*D4(K)+T5*D5(K)
99        M=M+1
```

The second version possesses another important advantage: if the array B is "virtual", the amout of paging is reduced. The inner loops of both versions are of course "vectorisable".

The number of parameters which can be refined by full-matrix least-squares on a conventional 16-bit mini-computer is restricted to about 100 by the limited memory. To retain the desirable convergence properties of full-matrix refinement without allowing the program to become "disk-I/O bound" the _blocked cascade algorithm_ may be used. This is a full-matrix refinement in which only part of the structure is refined each cycle. Structure factor contributions are only recalculated for those atoms which change in a given cycle, thereby saving computer time. Fixed atoms (e.g. H) are thus calculated only once per job. In the mini-computer program SHELXTL, the program decides on the basis of the shifts and esd's from previous cycles which atoms or rigid groups should be refined in the current cycle. In the mainframe SHELX-84, the user must specify how the atom list is divided up into "nodes", and the program decides which nodes to refine in which cycle. If bond length and other restraints are being used, it is thus possible to ensure that the atoms involved are usually in the same node. It is also possible to specify that for particular nodes, only positional parameters are refined on the even numbered cycles, and only thermal and occupation parameters on the odd numbered cycles.

The time required for convergence of the blocked cascade algorithm increases less rapidly with the number of atoms than for full-matrix refinement, so for large structures there is likely to be an optimum matrix size, which may be less than that indicated by the amount of memory available, especially for virtual computers.

References

Gilmore, C.J. (1984). _J. Appl. Cryst._ **17**, 42-46.

Konnert, J.H. and Hendrickson, W.A. (1980). _Acta Cryst._ **A36**, 344-350.

Mackay, A.L. (1984). _Acta Cryst._ **A40**, 165-166.

SHELX-84

By George M. Sheldrick,
Institut für Anorganische Chemie der Universität, Tammannstraße 4,
D-3400 Göttingen, Federal Republic of Germany

Introduction

SHELX-84 is the successor to the widely used SHELX-76 for the solution and refinement of crystal structures at atomic resolution from diffraction data. It is written in the same inimitable style as SHELX-76, and may be run on a wide variety of 32(or more)-bit computers without significant alteration. All routines are valid for all space groups, in conventional settings or otherwise. Storage is allocated _dynamically_ using a large single dimension array which resides in blank COMMON, so that there are _no restrictions_ on the _number_ of reflections, phase relations, scattering factor types, atoms, constraints, restraints, twin components, Fourier peaks, etc. This technique is particularly suitable for virtual memory machines, but is _not compatible_ with most _16-bit mini-computers_. The general SHELX-76 user interface, with free format input of crystal data, atoms etc. from the same file, has been retained. All SHELX-76 and SHELXTL reflection data formats (and a few others) can be read by SHELX-84, and in general only minor alterations (primarily for new commands) are needed to the instruction/atom files. So that the program remains small enough to run without overlay or complicated macros on most computers, it has been divided into two parts of about 6000 FORTRAN statements each, each of which may be run independently as a "stand-alone" program. The first part, SHELXS, is concerned with structure solution, the second, SHELXL, with refinement. Both programs include the necessary routines for processing instructions and reflection data, and both include the appropriate Fourier and unique molecule assembly facilities. They have been designed to be easy to use, with extensive use of default settings which may be changed, if necessary, by experienced users.

Changes from SHELX-76 and SHELXTL

Most of the changes take the form of extra instructions, but experienced SHELX users should note the following points before trying to use SHELXS-84 or SHELXL-84. The free-format input interpretation has been altered so that all characters after "/" or "=" (continuation marker) are ignored, so may be used as comments. However all other character strings (except the first four characters on a line, which as before form the instruction or atom name) which do not begin with a digit (or "-", "," or ".") are now interpreted as "labels". E.g. CA2+ could be used as a label on an SFAC instruction. These labels provide a very convenient method for global generation of appropriate restraints, hydrogen atoms, etc. Atom names, labels on atom instructions, and atom names may in general be used interchangedly. Thus the instructions (for SHELXL):

```
HATS 14 N
HATS 12 NH2 N31
HATS 0 C
```

would generate hydrogens of type 12 (planar $-NH_2$) on N31 and on all atoms which carry the label "NH2", and hydrogens of type 14 (planar -NH-) on all other atoms with a nitrogen scattering factor; the third HATS instruction gives the program discretion to generate appropriate hydrogens on all carbons. Similarly, the following instructions could be used to restrain a disordered PF_6^- group, in which the two sets of fluorines carry the labels FA and FB on each atom instruction (the phosphorus atom is common to both):

```
REST 31 .01 P4 F
RANG 90 .1 FA P4 FA
RANG 90 .1 FB P4 FB
```

where free variable 3 is the mean P-F distance.

Title, cell, symmetry, scattering factors and unit-cell contents are read almost exactly as in SHELX-76 or SHELXTL, except that the symmetry information is checked for internal consistency, and that explicit scattering factors must include the atomic weight, since the density is calculated. Scattering factors etc. are stored in the program for 32 common elements.

There may be only one HKLF instruction to read reflection data, and it must come last in the instruction file. The default OMIT level is F>4sigma(F), not zero. Users who use the common-reflection interlayer scaling facility (e.g. for Weissenberg data) or the face-indexed absorption correction procedures in SHELX-76 should retain their old program, because these facilities are not supported in SHELX-84.

The asymmetric unit of Fourier space may be generated automatically by the program, and the PLAN output includes bond angles and a shell of nearest neighbours; for large structures it may be split up into several lineprinter plots using the MOLE instruction.

SHELXS-84: Crystal Structure Solution

SHELXS provides facilites for direct methods, partial structure expansion and heavy-atom Patterson interpretation. The route taken by the program depends on whether the instructions TREF, TEXP or PATT are present; these cause appropriate defaults to be set up for FMAP, GRID and PLAN. For simple problems, the instructions will consist only of TITL...UNIT, PATT (interpret Patterson) or TREF (direct methods), and HKLF.

Generation of E-Values

E-normalisation is performed automatically, but may be controlled using the five parameters on the ESEL instruction; note that "ESEL -1.2" causes all calculations to be performed as triclinic non-centrosymmetric. The printed E-statistics have been considerably improved (e.g. by omitting the reflections at low and high 2-theta).

Patterson Interpretation

The algorithms used in Patterson interpretation are described in my Mülheim lecture on algorithms. The PATT instruction specifies the number of heavy atoms to be used to set up the "crossword table", the maximum number of vectors to be used for this table, the minimum heavy-atom to heavy-atom distance (default 1.2, but should be increased to about 6 for proteins), and the smallest peak height to be used. If atoms are also given, they are used directly to set up the crossword table. Alternatively, "PATT -200" may be used to set up a Patterson input file for the program PATSEE.

Partial Structure Expansion

This requires trial atoms (possibly from PATSEE: SPIN, MOVE and FRAG are allowed) and the TEXP instruction, which controls the tangent expansion. FMAP specifies the number of cycles of E-Fourier recycling which follows (default is two cycles).

Direct Methods

A multiple-random-start single-solution method is employed, which is usually followed by one cycle of E-Fourier recycling to improve the "best" E-map. This also gives an R-index which is a very good indication as to whether the solution is correct or not. TREF sets the number of phase permutations, the number of E-values (usually chosen automatically by the program), and the amount of output to be generated. TREF -n allows permutation number n to be evaluated, should the solution with the best CFOM not prove to be correct. If a large number of permutations is requested, the TIME instruction may be used to set the number of seconds after which the program should cease phase refinement and go on to the E-map etc. The NQAR instruction controls the use of the negative quartets; the default setting is that they are employed actively in the phase refinement only for symmorphic space groups. New phases are calculated using:

$$\text{phi} = \text{phase of } [\text{sigma}(\underline{E}_1 * \underline{E}_2) - \text{wn} * \text{t} * \text{sigma}(\underline{E}_3 * \underline{E}_4 * \underline{E}_5)]$$

where the first summation is over triple phase relations, the second over negative quartets. t is a constant (approximately $2/N^{1/2}$, where N is the number of "equal" atoms per lattice point) so that wn=1 is to a first approximation statistically correct (wn is set by NQAR). If wn is zero, negative quartets are used for NQUAL but not otherwise in the refinement. The new figure of merit NQUAL is more selective than NQEST:

$$\text{NQUAL} = \frac{\text{sigma}[\text{sigma}(\underline{E}_1 * \underline{E}_2) . \text{sigma}(\underline{E}_3 * \underline{E}_4 * \underline{E}_5)]}{\text{sigma}[|\text{sigma}(\underline{E}_1 * \underline{E}_2)| * |\text{sigma}(\underline{E}_3 * \underline{E}_4 * \underline{E}_5)|]}$$

where the outer summations are performed over all refined reflections, and the inner summations over triplets and negative quartet relations involving a given reflection, as above. Note that NQUAL approaches -1 for the correct solution, +1 for the uranium atom solution, and zero for random phases. However when both triplets (which, since the program employs Hull-Irwin weights, drive R(alpha) to a minimum) and negative quartets (which drive NQUAL to -1) are employed in the refinement, the "best" solution must have good values of both R(alpha) and NQUAL. The combined figure of merit reflects this, and is valid under a wide range of conditions:

$$CFOM = R(alpha) / (1-wq*NQUAL)^2$$

where wq usually takes a value close to 1, but is reduced by the program when there is only a small number of negative quartets.

Proteins

Both the Patterson interpretation and the direct methods perform well in locating heavy atoms from isomorphous delta-F data. It is necessary to set the cell contents (UNIT) to the correct number of heavy atoms plus the square root of the number of C, N and O atoms (assigned SFAC N). Negative quartets should <u>not</u> be used in direct methods, and FMAP 4 (no recycling) and an appropriately small number of peaks (PLAN) must be specified (the default value depends on the unit-cell volume !).

SHELXL-84: Crystal Structure Refinement

The least-squares refinement part of SHELX-84 is at a much less advanced stage of testing than the structure solving part, and so the following information is provisional. Important new facilities compared with SHELX-76 include refinement of data from twinned crystals (Jameson, 1982), and refinement of a "racemic twinning" parameter (Flack, 1983) as an aid to absolute structure assignment. Empirical absorption corrections are applied by the method of Walker and Stuart (1983), which is incorporated into the least-squares refinement. Refinement on F or F^2 is allowed, using a flexible blocked-cascade algorithm (see lecture on algorithms) which reverts to full-matrix for small problems. The weighting scheme may be set for high-angle refinement, and the isotropic extinction correction has been improved.

There are improved and much extended algorithms for hydrogen atom generation, taking symmetry, disorder, and possible hydrogen bonds into account. Hydrogen contributions may be calculated once only (HATS) or included in the refinement in the form of a riding model or rigid groups (AFIX). General rigid groups may be fitted using literature or force-field coordinates, and linked rigid groups are permitted. A new 4-parameter rigid group model is convenient for methyl groups, and restraints may be applied to any atom in a rigid group.

In addition to conventional isotropic and anisotropic atoms, 2-parameter thermal ellipsoids have also been introduced (see lecture on algorithms). U_{ij}-constraints may be generated automatically for atoms on special positions. Additional restraints include bond angles, isotropy of temperature factors (to prevent them going n.p.d. !), and linear sums of free variables (for multiple site occupancy etc.). The use of atom names, global labels, and scattering factor names on the restraint and HATS instructions considerably simplifies the user interface. If the user has forgotten to fix a floating axis, the program automatically generates an appropriate restraint.

A results file is produced after every cycle (in case the computer crashes during a long job !), and the TIME instruction may be used to terminate refinement and tidy up before the CPU time allocation expires. A checklist and tables are provided in suitable format for Acta Cryst. or deposition. The appropriate subroutine can easily be converted by the user to an (interactive) stand-alone program for more sophisticated production of tables for publication.

References

Flack, H.D. (1983). *Acta Cryst*. **A39**, 876-881.

Jameson, G.B. (1982). *Acta Cryst*. **A38**, 817-820.

Walker N. and Stuart, D. (1983). *Acta Cryst*. **A39**, 158-166.

HIGHER INVARIANTS

by H. Schenk

Laboratory for Crystallography, University of Amsterdam,
Nieuwe Achtergracht 166, 1018 WV Amsterdam, The Netherlands.

Introduction

In present direct method procedures higher invariants play a more and more significant role. They serve in nearly all stages of the phase determination from the starting set procedures to the determination of the correct solution by means of merit and also in extension and refinement processes.

Direct methods started in 1948 when Harker and Kasper published their inequalities and Gillis used these to solve a structure. Around 1950 many theoretical developments were invented (e.g. Karle and Hauptman, 1950; Sayre, 1952; Hauptman and Karle, 1953; Kitaiporodsky, 1954). The first useful approach to higher invariants came 20 years later when Schenk (1973) introduced a way in using the seven |E| magnitudes of the reflections H, K, L, -H-K-L, H+K, H+L and K+L to estimate the quartet phase sum

$$\Phi_4 = \phi_H + \phi_K + \phi_L + \phi_{-H-K-L}$$

Because the indices of the four main reflections H,K,L and -H-K-L sum to zero it is a structure invariant, and for the reason that it involves more than three reflections it is called a higher invariant. In the first paper on quartets (Schenk, 1973) it was shown that when the cross terms H+K, H+L and K+L are used in addition to the main terms, more reliable estimates of Φ_4 are obtained. Also a procedure was presented in which these quartets were used successfully in determining starting sets within the group of strongest reflections.

This paper was the starting point for fruitful research on higher invariants (such as quartets and quintets) and seminvariants. The latter are generally estimated via higher invariants.

It is the purpose of this paper to extend understanding of these higher invariants and their applications. Therefore sections are included which deal with the general scope and properties of the invariants, with the theory behind the invariants and with realized applications. The latter are not dealt with extensively, because recent reviews already cover this material sufficiently (e.g. Schenk and Kiers, 1984).

The physical meaning of the phase relationships

A strong structure factor F_H implies that the term H in the Fourier summation of the electron density

$$\rho(r) = \Sigma |F_H| \cos 2\Pi(H.r + \phi_{hkl}) \quad (1)$$

is an important one, giving rise to a substantial contribution to the image of the electron density at the positions of many of the atoms. As has ben described previously (Schenk, 1973; Schenk, 1974; Schenk , 1981, Schenk, 1984 a,b) the capability of single structure factors to add substantially to $\rho(r)$ at the positions of the atoms, leads to simple physical explanations of Σ_1-relation, the triplet relation

$$\phi_H + \phi_K + \phi_L \simeq 0, \quad (2)$$

a special negative quartet relation

$$2\,\phi_H + \phi_{-H+K} + \phi_{-H-K} \simeq \pi, \quad (3)$$

the positive quartet relation

$$\phi_H + \phi_K + \phi_L + \phi_{-H-K-L} \simeq 0, \quad (4)$$

and the negative quartet relation

$$\phi_H + \phi_K + \phi_L + \phi_{-H-K-L} \simeq \pi. \quad (5)$$

These explanations make use of the fact that it is likely that a number of strong intensities contribute to the electron density of the same atoms. This leads to a complete description of the relationships with in the case of expressions 3, 4 and 5, the proper role of the cross terms. For the Σ_1 relation in space group $P\bar{1}$ it is shown that when the structure factor magnitudes of the reflections H and 2H are both strong the phase of 2H most likely is being 0. In the case of the triplet relationship it appears that when all three reflections involved are strong the sum of the three phases involved is most likely being 0 (relation 2). In a similar way it follows that relation 3 is likely to be true for strong reflections H, H+K and H-K, and weak reflections K and 2H. For the two quartet relations 4 and 5 it is shown that they are more probably true for large $|E_H|$, $|E_K|$, $|E_L|$ and $|E_{-H-K-L}|$ in both cases and for the cross terms $|E_{H+K}|$, $|E_{H+L}|$ and $|E_{K+L}|$ being large and small for 4 and 5 respectively.
Because in the referred papers these relationships are dealt with in great

detail on the basis of electron density considerations, we will suffice here with the above short descriptions.

Structure invariants and structure seminvariants

Examples of structure invariants and structure seminvariants are already given in the previous paragraph. First the definitions are given:

- a *structure invariant* is a quantity which remains unchanged when the origin is shifted to another position in the unit cell.
- a *structure seminvariant* is a quantity which remains unchanged when the origin is shifted from the one origin, as given for a particular setting of a particular spacegroup by International Tables for Crystallography, to the other.

From the definition of structure invariants it follows that they have the general form

$$\sum_{i=1}^{n} q_{H_i} \quad \text{with} \quad \sum_{i=1}^{n} H_i = 0 \tag{6}$$

in which q can be replaced by for instance E, F or Φ.
Intensities are measured and consequently they must be structure invariants in the sense of the definition. This follows also from expression 6 by choosing n=2 and q=F. With n=3 and q=Φ expression 6 reduces to the triplet invariant and with n=4 and q=Φ to the quartet invariant.
It must be strongly emphasized that although all quantities meeting the requirements of expression 6 are invariants, they are only useful for structure determinations when their sum

$$S = \sum_{i=1}^{n} q_{H_i} \tag{7}$$

can be estimated and that is only the case under special conditions. For instance, for triplet invariants they are only useful in case that the three |E|'s involved are strong and, as a result, the most likely value of S is zero. It is therefore sensible to make a distinction between invariants and phase relationships following from invariants. In the preceding paragraph we have seen a number of these phase relationships and it was stated under which conditions they most probably hold.
From the above it will be obvious that the term "structure invariant" is confusing, because expression (6) does describe a quantity which is invariant under origin shifts, and is not invariant with respect to the

structure at all. Only under special conditions an invariant (6) may carry structural (phase) information and then turns into a useful relationship. Thus, its name is never judged by its properties and a name like "origin invariant" would have been more appropriate.
Also structure seminvariants are invariant with respect to an origin shift; again a confusing name. Now they follow from the conventional choices of origin as they are given for all space groups in the International Tables for Crystallography. For instance in space group $P\bar{1}$, the origin is conventionally chosen at one of the 8 different centres of symmetry. When the origin is shifted from one of these centres to any other of them it will appear that the phase of a reflection 2H, e.g. a reflection with ggg indices, remains unchanged. Thus in $P\bar{1}$ and any other centrosymmetric spacegroup the reflections 2H are seminvariants. In general for these spacegroups all quantities

$$\sum_{i=1}^{n} q_{H_i} \quad \text{with} \quad \sum_{i=1}^{n} H_i = 0(\text{modulo } 2,2,2) \tag{8}$$

are seminvariants, in which 0(mod 2,2,2) implies that the sum of the indices is ggg. In all spacegroups with exception of P1, seminvariants are present, sometimes following (8) but also following other rules.

Estimating the phase sums of invariants and seminvariants

In previous paragraphs some phase relationships were mentioned already and under what conditions they can be reliably estimated. Phase relationships based on higher invariants such as quartets heavily depend on the use of cross terms, and the same is true for relations based on seminvariants.
In recent years the knowledge which structure factor magnitudes play a role in the estimation of phase sums of invariants and seminvariants has been independently systemized by Hauptman (1977, 1978) and Giacovazzo (1977, 1980) resulting in their respective Neighborhood and Representation theories. Both theories differ only in less extent from each other and are based on three well established principles, namely the cross term principle, the identity principle and the representation or extension principle. Then additionally Hauptman makes implicit use of symmetry whereas Giacovazzo introduces symmetry in a more elegant and explicit way. In this paper a review of the theories would occupy too much space, and instead, a brief summary of the three underlying principles will be given.
As already mentioned, cross terms are essential for the estimation of

higher invariants. Cross terms were used for the first time by Schenk (1973 a,b) and Schenk and De Jong (1973) for quartets and special quartets, and by Schenk (1975) and Schenk and Van der Putten (1977) for quintets. They originate from the fact that higher invariants can be obtained directly

$$\phi_n = \phi_H + \phi_K + \phi_L + \phi_M + \ldots. \tag{9}$$

but also by summing suitable lower order invariants

$$\phi_n = (\phi_H + \phi_K + \phi_{-H-K}) + (\phi_{H+K}) + \phi_L + \phi_M + \ldots.) \tag{10}$$

In the latter case the E-magnitude of reflection H+K plays also an important role, and a term like this is called a cross term. All cross terms of expression (9) are H+K, H+L, ..., K+L, ..., H+K+L, ..., etc. The value of ϕ_n is controlled by both the $|E|$'s of the main terms H, K, etc. and those of the cross terms.

The identity principle is a generalization of the quadrupoles of triplets. Quadrupoles were implicitely introduced by Sayre (1952) and later worked out by e.g. Vaughan (1956), Karle and Hauptman (1957), De Vries (1963) and Viterbo (1976). A quadrupole consists of 4 triplet relationships:

$$\begin{aligned} \phi_H + \phi_K + \phi_{-H-K} &= \phi_3 \\ \phi_P + \phi_{-H} + \phi_{H-P} &= b \\ \phi_{-P} + \phi_{-K} + \phi_{K+P} &= c \\ \phi_{H+K} + \phi_{P-H} + \phi_{-P-K} &= d \end{aligned} \tag{11}$$

with the identity

$$\phi_3 + b + c + d = 0 \tag{12}$$

From (12) it follows that ϕ_3 is also dependent on the phase sums b, c and d from three other triplet phase relations. Therefore the value of ϕ_3 is not only controlled by $|E_H|$, $|E_K|$ and $|E_{-H-K}|$ but also by the $|E|$'s of the reflections P, H-P and K+P, where P ranges over all reciprocal space.

The quadropole is just one example of an identity and in a similar way many identities can be set up for any invariant, always including at least one free reciprocal vector P. Therefore calculations of phase relationships based on identities are always computer-time-demanding.

The last important ingredient of the theories of Giacovazzo and Hauptman is the representation or extension principle. This principle is simply given by the expression

$$\phi_n + \phi_p = \phi_m \tag{13}$$

in which ϕ_n is the phase relationship the value of which is being estimated,

ϕ_p is a phase relationship with a known value and ϕ_m is a phase relation of a higher invariant, which can be estimated using cross terms and (or) identities. Relation ϕ_n may be either an invariant or a seminvariant and as a result ϕ_p is also either an invariant or seminvariant respectively. To be more explicit, the oldest known example of (13) in space group P1 is given by

$$\phi_{2H} + \phi_{-H} + \phi_{-H} = \phi_3 \qquad (14)$$

in which the value of ϕ_{2H} is calculated using the triplet relationship ϕ_3 and the fact that $\phi_{-H} + \phi_{-H} = 0$ in centrosymmetric space groups. This is the so called Σ_1 relation and as such already present in the work of Harker and Kasper (1948) and Hauptman and Karle (1953). All other Σ_1-relationships in space groups of higher symmetry obey relation (13), with for ϕ_p in many cases more than one possibility, so that ϕ_n is built up from a number of contributors. Giacovazzo and Hauptman have generalized this principle, so that for instance also the following example is included:

$$\phi_H + \phi_K + \phi_{-H-K} + \phi_P + \phi_{-P} = \phi_5 \qquad (15)$$

in which the triplet phase sum is calculated from the quintet ϕ_5, taking into account that $\phi_P + \phi_{-P}$ is trivially equal to zero. The quintet (and thus the triplet!) can be estimated using the main terms H, K, -H-K, P and -P and the cross terms H+K, H, K, O, ±P+H, ±P+K, and ±P-H-K. It can be noted that this way of looking at the triplet relation has similarities with the identity (12).

Using these three principles the important reflections for the estimation of any invariant or seminvariant can be traced, and moreover, a qualitative prediction can be made about their influence on the invariant or seminvariant under study by making use of analogies. E.g. relation (15) will have $\phi_3 = \phi_5 \simeq 0$ for all main and cross terms of the quintets large.

Probability distributions for phase relationships

Once the reflections controlling the value of the phase sum of a (sem)-invariant are identified, a conditional joint probability distribution can be derived, describing the probability that the phase sum has a certain value, given the |E|'s of all reflections involved. Throughout the years many scientists have been contributing to this part of direct methods, e.g. Hauptman, Karle, Cochran, Kitaigorodsky, Klug, Bertaut, Naya, Giacovazzo, Heinerman, and recently Bricogne and Peschar.

Starting point of the derivation of a joint probability distribution is a process to mimic structures. Then given such a structure the formalism of probability theory is used to derive a joint probability distribution. Finally the conditional joint probability distribution is found by precizing some parameters of the joint probability distribution.

For the process of mimicing structures in general randomly chosen coordinates are given to all atoms. Thus for all positions in the unit cell there is an equal chance to find an atom. Of course, for real structures this is an incorrect assumption, because all atoms are situated at sites which satisfy chemical and physical rules like chemical bonding and Van der Waals forces. Inclusion of these rules in the process of mimicing structures complicates the procedure to arrive at a conditional joint probability distribution enormously.

In the second step probability theory is used. However, in general approximations are applied in order to make the derivation feasible. Different authors use different approximations and may thus arrive at different joint probability distributions for the same phase relationship.

This paper is not intending to give a full account of theoretical direct methods. In stead, some recent results in the field of quartet relations are compared. After the first empirical observations concerning 7 magnitude quartets (Schenk, 1973a) theories were developed by Hauptman (1975) and Giacovazzo (1976) along the lines described above. The conditional probability distribution of Giacovazzo (1976) is given by

$$P(\phi_4,R_1,R_2,R_3,R_4,R_5,R_6,R_7) = C^{-1}\exp[-2N^{-1}R_1R_2R_3R_4(R_5^2-1)(R_6^2-1)(R_7^2-1)] \qquad (16)$$

and the one of Hauptman (1975) by

$$P(\phi_4,R_1,\ \ldots,R_7) = C^{-1}\exp[-2N^{-1}R_1R_2R_3R_4\cos\phi_4] \times \times I_0(2N^{-1/2}R_5Z_5)I_0(2N^{-1/2}R_6Z_6)I_0(2N^{-1/2}R_7Z_7) \qquad (17)$$

with $Z_5 = [R_1{}^2R_2{}^2 + R_3{}^2R_4{}^2 + 2R_1R_2R_3R_4\cos\phi_4]^{1/2}$

and similar expressions for Z_6 and Z_7. In all expressions R_1 is the random variable associated to $|E_H|$, R_2 to $|E_K|$, R_3 to $|E_L|$, R_4 to $|E_{-H-K-L}|$, R_5 to $|E_{H+K}|$, R_6 to $|E_{H+L}|$, and R_7 to $|E_{K+L}|$. The form of expression (16) shows that the maximum of P is either found at ϕ_4=0 or at $\phi_4=\pi$, according to the values of R_5, R_6 and R_7. For large values of these cross terms ϕ_4=0 and for small values the sign of the exponential part changes and consequently $\phi_4=\pi$.

The formula of Hauptman, however, is capable to predict all values in the range $0 \leqslant \phi_4 \leqslant \pi$, and has therefore a wider range of possible applications.

Nevertheless, also this expression has shortcomings. In particular the fact, that in real structures the predicted Φ_4-values show systematic errors for both the modes and the means of the conditional probability distributions, is a limiting factor for successful applications.

The Giacovazzo- and Hauptman-expressions have been derived neglecting higher order terms. In order to judge whether this leads to the systematic errors of the Φ_4 estimates in my laboratory Peschar (1984) followed another route to the joint probability distribution. This route enables the use of higher order terms to any specified order of N. Therefore the characteristic function C of the joint probability distribution (jpd) is brought into a special form and then expanded in a Taylor series. The result can be Fourier-transformed term by term. In the final form of the jpd products of the Hermte and Laguerre polynomials play an important role. In order to make this process feasible a computer program has been written, which does all the hard work from both the mathematical and administrative point of view. The program derives the expression for the conditional jpd to any specified order of N and delivers the result as a series of terms on a disk file. Typically the conditional jpd correct to the order of N^{-3} contains 1010 terms and that correct to N^{-4} 9133 terms. These numbers show the impossibility to derive the expressions by hand.
The fact that the pd is available as a computer disk file makes it easy to test it, which has been done extensively.

A first test shows that convergence is normally reaced using the N^{-3} expression. Only when all reflections involved are very strong the N^{-4} expression has to be used in order to ensure convergence.

Another test shows that the estimates of Φ_4 do not contain any systematic error with respect to the true phase sums, when the mean values of the cjpd's in the N^{-3} form are used, whereas the systematic errors in the case of Hauptman's formula are as big as 15° on average. Moreover, there is no systematic error variation over the whole range of $0 \leq \Phi_4 \leq \pi$ whereas the differences in the case of expression (17) vary from 20 to 10° for different Φ_4 averages.

A last advantage of the new formula is its lower error level; the absolute errors of the estimated Φ_4 with respect to the true Φ_4 are reduced by approximately 15%. In conclusion it has been shown from this investigation that the incorporation of higher order terms give rise to superior jpd's for the quartet phase relationship.

Along the same route, taking advantage of similar computer programs, also other pd's are derived with similar results which will be published in due course (Peschar, 1984) and will be presented partly in Hamburg (Peschar and Schenk, 1984).

Application of higher invariants in practical Direct Methods

The applications of higher invariants in practical direct methods vary from well-established widely used ones to applications which have not been developed yet beyond the research stage.
Among the well-established applications are the use of quartets in the starting set determination and the use of figures of merit based on quartets, quintets and seminvariants. These applications are implemented in several Direct Method programs, of course including our own SIMPEL program (see these proceedings). For an overview of the different Direct Method programs see the paper of K. Huml in these proceedings. Elsewhere (Schenk and Kiers, 1984) the applications themselves are described in more detail.
In the same paper also the less open applications are dealt with in some detail. A number of them make use of the fact that for many invariants and seminvariants absolute values of sums of phases can be estimated which differ from 0 and π. These estimates are therefore enentiomorph sensitive. It would of course be better if they were enantiomorph specific, that is to say that Φ_n itself is known and not its absolute value only. In practice the latter is mostly the case, so that procedures based on these relationships must have an inbuilt routine which determines whether the sign of the invariant is plus or minus. It would be helpful, in particular in space groups like $P2_1$ and P1, to be able to use standard methods based on $|\Phi_n|$-values to overcome the enantiomorph problems. However, they are still not grown beyond their childhood and as such only useful at the places where they were invented.

Conclusion

In conclusion the field of higher invariants has been enriched by the possibilities of Direct Methods substantially. Moreover, since not all potential applications are yet available in the form of user programs and secondly since research in this field is still proceeding, the future Direct Methods procedure will gain more from higher invariants.

References

De Vries, A. (1963). Thesis, Utrecht.

Giacovazzo, C, (1976). Acta Cryst. A32, 91.

Giacovazzo, C. (1977). Acta Cryst. A33, 933.

Giacovazzo, C. (1980). Acta Cryst. A36, 362.

Gillis, J. (1948). Acta Cryst. 1, 174

Harker, D and Kasper, J.S. (1948). Acta Cryst. 1, 70.

Hauptman, H. (1975). Acta Cryst. A31, 680.

Hauptman, H. (1977). Acta Cryst. A33, 553.

Hauptman, H. (1978). Acta Cryst. A34, 525.

Hauptman, H. and Karle, J. (1953). A.C.A. Monograph No. 3. Pittsburgh, Polycrystal.

Karle, J. and Hauptman, H. (1950). Acta Cryst. 3, 181.

Karle, J. and Hauptman, H. (1957). Acta Cryst. 10, 515.

Kitaigorodsky, A.I. (1954). Dokl. Acad. Nauk. SSSR 94, 225.

Peschar, R. (1984). In preparation.

Peschar, R. and Schenk, H. (1984). Acta Cryst. A40, C426.

Sayre, D. (1952). Acta Cryst. 5, 60.

Schenk, H. (1973). Acta Cryst. A29, 77.

Schenk, H. and de Jong, J.G.H. (1973). Acta Cryst. A29, 31.

Schenk, H. (1973b). Acta Cryst. A29, 480.

Schenk, H. (1975). Acta Cryst. A31, S14.

Schenk, H. and van der Putten, N. (1977). Acta Cryst. A33, 368.

Schenk, H. (1979). J. Chem. Ed. 56, 383.

Schenk, H. (1981). Acta Cryst. A37, 573.

Schenk, H. (1984a) in Methods and Applications in Crystallographic Computing, edited by S.R. Hall and T. Ashida, Clarendon Press, Oxford, p. 82.

Schenk, H. (1984b). An Introduction to Direct Methods, IUCr pamphlet, University College Cardiff Press, Cardiff.

Schenk, H. and Kiers, C.T. (1984). In Methods and Applications in Crystallographic Computing, edited by S.R. Hall and T. Ashida, Clarendon Press, Oxford, p. 96.

Vaughan, P.A. (1956). ACA Annual Meeting, Frenchlick, Ind.

Viterbo, D. (1976). In Direct Methods in Crystallography, edited by H. Hauptman, Buffalo, p. 117.

SIMPEL 83, A PROGRAM SYSTEM FOR DIRECT METHODS

By H. Schenk and C.T. Kiers

Laboratory for Crystallography, University of Amsterdam,
Nieuwe Achtergracht 166, 1018 WV Amsterdam, The Netherlands.

The program system SIMPEL is a complete direct methods system, which may be entered with |F|-values and may produce an E-map of the structure. The unique part of SIMPEL is formed by a series of routines, successively gathering the relationships, determining a starting set, carrying out a symbolic addition, finding the correct numerical values for the symbols and finally executing a numerical phase extension and refinement. The other parts of the program system may consist of any normalization program and any Fourier program, peak search and map interpretation program. Usually these parts are the same as those used in MULTAN, however, they may also belong to other systems, such as the XRAY or XTAL system or a local system. In our own set-up it is a mixture of programs of different origin. The end-point in our system is a very fast diagonal least squares structure factor program, which enables us to check interactively solutions from E-maps and refine structures for minimal computing-costs from R=45% down to R=15%.
This paper will contain short descriptions of the unique parts of SIMPEL only.

Collecting phase relationships

The different phase relationships are all collected by means of similar algorythms. These algorythms are based on the use of two arrays, one (array A) in which all strong reflections, the phases of which are aimed at, are taken up and another (array B) in which the information of all reflections is stored three dimensionally on the basis of the crystallographic indices h, k and l.
All relations are searched for by taking the relevant reflections from array A, calculating indices for the remaining reflections and looking those up in array B. For instance quartets are gathered by taking H, K and L from array A, calculating -H-K-L and looking this reflection up in array B. If it is a quartet with a quartet product $N^{-1}|E_H E_K E_L E_{-H-K-L}|$ above a certain limit, also the information about the cross terms H+K, H+L and K+L is looked up in array B and the quartet with all its information is stored.

Our program searches for triplets, quartets (negative, positive and enantiomorph sensitive ones), two-dimensional quartets (H,H,H+K and H-K), and Σ_1-relationships. It is very straightforward to add routines for calculating other relationships such as quintets and in our research version of SIMPEL quintet and seminvariant routines are present.
The algorythms have been described in more detail elsewhere (Overbeek and Schenk, 1984)

The starting set procedure

The determination of the starting set begins with a convergence procedure similar to that devised by Germain, Main and Woolfson (1970). This procedure searches for that reflection which is weakest linked to the other reflections by means of the phase relations and then this reflection is removed from the set of reflections. At the same time all its phase relations are removed from the collection of phase relationships. This process is repeated until no reflections are left in the set. Any time a reflection is removed without having phase relationships linking it with other phases, this reflection has to be taken up in the starting set.
Our procedure uses triplets and seven magnitude quartets for a relative small number of strong reflections; usually the number is in between 30 and 60. This procedure goes back to the first paper on seven magnitude quartets (Schenk, 1973) when we discovered that most of the strong quartets are found within the group of very strongest reflections. Since in a starting set reflections with large |E| are preferred, the use of triplets and quartets generally leads to very good starting sets. After completing the convergence procedure the origin is fixed and to all other starting set reflections symbolic phases are attached.
The second step in our procedure is a few cycles of symbolic addition, using again quartets and triplets, employing very strict acceptance criteria and not allowing relations among the symbolic phases. This leads to a much larger starting set. Then finally in the third step this set is subjected to an accessibility test, in which it is checked, without using symbols, whether the phases of al reflections can be evaluated from the starting set using the adopted strict accepting scheme. If this is not the case, a warning is given because weak links between phases are found and an additional symbolic starting reflection may be neccessary. In next versions this will be done by the system itself, now this is left to the user. As a rule, however, the starting sets produced by SIMPEL are quite satisfactory, and no additional starting phases are necessary.

When SIMPEL does not provide the solution of the phase problem this could be caused by a bad starting set. In that case one can very easily develop an own starting set by using a triplet list produced by SIMPEL and feed this starting set to the symbolic addition procedure. Facilities for doing this are present in the system.

The symbolic addition procedure

The next step in the process is the extension of the starting set. This is done in SIMPEL using triplet relations only. It is very essential that no errors are included during this process and therefore the criteria used to accept a symbolic phase for a reflection are very strict, in particular in the beginning. In general, no single indication will be accepted unless it belongs to the ten to fifteen strongest triplet relationships. For multiple indications giving rise to the same symbolic phase, high acceptance limits are applied and when a reflection gets more than one different symbolic phase via its triplets, this reflection is neglected at all in the further procedure. In general these precautions make sure that the resulting set of symbolic phases contain the correct solution, that is to say that when the correct values for the symbols are substituted the phases of the reflections are good enough to image the structure. In the centrosymmetric case about 20 cycles of extension are carried out and in the noncentrosymmetric case 10. In the end a summary has been given of all used and unused information. From those reflections having more than one different phase indication and neglected in the symbolic addition itself, now relations between the symbols are collected and stored to use in the successive step.

Numerical values for the symbols: Figures of Merit

With the extended group of phased reflections as input numerical values for the symbols are calculated using a number of Figures of Merit (FOM's). The general tactics to calculate FOM's is first to substitute the symbolic phases into the relationships of the FOM, then to combine all identical symbolic phase sums by summing their weights and finally calculating the FOM's by replacing the symbols by numerical trial values.

In SIMPEL several FOM's may be used:

1) The Σ_1-consistency FOM Q (Schenk, 1971), which can be calculated very fast;
2) The Positive Quartet Criterion (PQC, Schenk and Kiers 1984);

3) The Negative Quartet Criterion (NQC, Schenk, 1974, with weights from Schenk, 1975);
4) The Σ_1-criterion (Overbeek and Schenk, 1976);
5) The Harker-Kasper Criterion (HKC, Schenk and De Jong, 1973, Schenk, 1973b).

The FOM's 2, 3 and 5 are based on higher invariants. In noncentrosymmetric spacegroups with centric projections the reflections with restricted phases are used to determine symbol values modulo π. Apart from the separate FOM's also a Combined FOM is calculated, which mostly discriminates the correct solution without difficulties.

In general there is few correlation between the different Figures of Merit, thus resulting in a strong Combined FOM. Recent results show that the Negative Quartet Criterion is correlated strongly with the in other systems used ψ_0 check. However, NQC has much more discriminating power than ψ_0, and has therefore to be used preferably.

For different space groups it is advisable to rely on groups of different FOM's. For instance, whereas in space groups like $P2_1/c$ the Q and PQC are superior, in space groups like $P\bar{1}$ they are unreliable as a result of the lack of translational symmetry. In this space group NQC and HKC are the ones which generally guide the phase determination to a successful end. The projection criterion is particularly successful in space groups like $P2_12_12_1$, in which many reflections have restricted phases.

Numerical phases for all strong reflections

After determining the symbols all symbolically phased reflections can get a numerical phase. These phases are then used as starting point for a numerical phase extension, because in general the number of phased reflections is yet unsufficient to calculate a good E-map. In the centrosymmetric case this is achieved by means of a fast Σ_2 refinement and extension, in the noncentrosymmetric case by means of a usual tangent refinement.

Implementation of SIMPEL

SIMPEL now exists in two modifications. The one is our CDC Cyber version, in which the 60 bit words are used to store as much information as possible by wordpacking. The other is the PDP11 version, which forms part of the ENRAF-NONIUS Structure Determination Package. Both systems are very similar if not identical and user-friendly. The PDP 11 version runs also in the Digital P350 microcomputer. Soon SIMPEL will also be implemented in the XTAL system.

Some notes about the performance of SIMPEL

Because the procedures of SIMPEL differ completely from all other existing direct methods systems like MULTAN, Mithril and Shelx, it is well possible that a failure of another system is solved in a default SIMPEL run. Of course, the reverse may also be the case. However, in any case SIMPEL will present its results in a relative short time compared with the other systems, because SIMPEL carries out only two phase extensions, one symbolical and one numerical extension, whereas all multisolution programs do "multi" extensions, in which "multi" is large and to be defined by the uses. Therefore people familiar with SIMPEL will try that system first, just because its speed.

The rate of success of SIMPEL in the space groups $P\bar{1}$ and $P2_1/c$ is equal and over 90%, provided that the appropriate FOM's are used. The same rate of success can also be claimed for $P2_12_12_1$, although the number of structures tried in this space group is smaller. In $P2_1$ the rate is less, however, this applies to other programs as well. A number of results is reviewed in Schenk (1983), and others will be reviewed in Overbeek (1985).

The investigations were in part supported by the Netherlands Foundation for Technical Research (STW) and in part by the Netherlands Foundation for Chemical Research (SON). The authors wish to thank their collegues N. van der Putten, G.J. Olthof, J.D. Schagen, R. Peschar, R. Driessen, K. Goubitz, and in particular A.R. Overbeek for their contributions to the research version of SIMPEL. Otherwise SIMPEL 83 could not have been written.

References

Germain, G., Main, P. and Woolfson, M.M. (1970). Acta Cryst. B26, 274.

Overbeek, A.R. (1985). Thesis in preparation.

Overbeek, A.R. and Schenk, H. (1976). Proc. K. Ned. Akad. Wet. B79, 341.

Schenk, H. (1971). Acta Cryst. B27, 2039.

Schenk, H. (1973a). Acta Cryst. A29, 77.

Schenk, H. (1973b). Acta Cryst. A29, 480.

Schenk, H. (1974). Acta Cryst. A30, 477

Schenk, H. (1975). Acta Cryst. A31, 257.

Schenk, H. (1983). Recl. Trav. Chim. Pays-Bas 102, 1.

Schenk, H. and de Jong, J.G.H. (1973). Acta Cryst. A29, 480.

Schenk, H. and Kiers, C.T. (1984) in Methods and Applications in Crystallographic Computing, edited by S.R. Hall and T. Ashida, Clarendon Press, Oxford, p 96-105.

Schenk, H. and Overbeek, A.R. (1984) in Methods and Applications in Crystallographic Computing, edited by S.R. Hall and T. Ashida, Clarendon Press, Oxford, p 91-95.

MULTAN - a program for the determination of crystal structures

Peter Main, Physics Department, University of York.
York YO1 5DD, England.

1. Summary.

Multan is a computer program which determines crystal structures from X-ray diffraction data by direct phase determination. It uses the tangent formula of Karle & Hauptman (1956) with a weighting scheme similar to that suggested by Hull & Irwin (1978). Starting values for unknown phases are given a number of different values, thus producing several plausible sets of phases. Such a scheme was first suggested by Germain & Woolfson (1968) and has since become known as the "multisolution method". The sets of phases are ranked in order by figures of merit and E-maps are calculated using the best of them. The maps are interpreted by the program and often it is the map from the highest ranked phase set which reveals the structure.

Structures consisting of up to 120 equal, independent atoms should be solved by Multan without too much difficulty. Structures having pseudo-translational symmetry can be dealt with according to the technique developed by Fan, Yao, Main & Woolfson (1983).

The program is written in standard Fortran IV, uses 16-bit integers and runs in 56 Kbytes of memory. It should therefore run on a modest minicomputer (or a good micro) provided that up to 2 Mbytes of disc space is available. It also works efficiently on larger machines. Copies of the program may be obtained from the author.

2. The multisolution method.

One of the features of the multisolution method, as opposed to symbolic addition, is that numerical values for phases are used throughout. This simplifies programming, especially for non-centrosymmetric structures, but it also means that the number of unknown phases to be assigned values at the beginning is not limited. (It is impractical to deal with hundreds of symbols.) Thus it is possible to assign values to all unknown phases at the beginning of phase determination and refine them using the tangent formula. With careful use of weights in the tangent formula, this is more powerful than determining phases systematically from a small number of initial phases.

The multisolution method is essentially a Monte Carlo approach to the phase problem. Sensible values can not be assigned to starting phases except by chance. It is therefore necessary to repeat the phase determination for many different sets of starting values. This appears to make the method very inefficient. However, a single-solution method which always works does not yet exist, so the multisolution method continues to be used. For non-centrosymmetric crystals, about 30 sets of phases should be sufficient to solve structures containing less than about 40 independent atoms. With more than 100 independent atoms, several hundred phase sets may be necessary.

3. Outline of the program.

The principal operations carried out by Multan are shown below in logical order. Brief descriptions of some of these operations are given in later sections of this paper.

a) calculation of E's from F_{obs} values.

Great care must be taken in the estimation of E's for low angle reflexions.

b) set up phase relationships.

Proper application of space group symmetry to phase relationships leads to more accurate phases.

c) find the reflexions to be used for phase determination.

Use the convergence procedure of Germain, Main & Woolfson (1970) to pick the strong reflexions to be used for phase determination.

d) phase determination.

Starting phases are assigned using magic integer phase permutation (Main, 1978). Either a small number of starting phases may be used or values may be assigned to all phases and then refined. In both cases, the weighted tangent formula is used.

e) calculate figures of merit.

The figures of merit can be used to detect a correct

solution and stop the phase determination at that point.

f) calculate and interpret E-map.

Maps are calculated in order of figure of merit. Simple stereochemical criteria are applied to the peaks found in order to produce a chemically sensible molecule. Alternative interpretations of the map may be produced.

4. Calculation of E's.

Normalised structure amplitudes (E's) may be defined by

$$|E(\underline{h})|^2 = \frac{I(\underline{h})}{\varepsilon_h \langle I \rangle} \qquad (1)$$

where $\langle I \rangle$ = expected intensity

and ε_h takes account of the effect of space group symmetry on $I(\underline{h})$.

The best way of estimating $\langle I \rangle$ is as a spherical average of the actual intensities. In practice, the reflexions are divided into ranges of $\sin^2\theta/\lambda^2$ and averages taken of intensity and $\sin^2\theta/\lambda^2$ in each range. Reflexion multiplicities and also the effects of space group symmetry on intensities must be taken into account when the averages are calculated. Sampling errors can be decreased at low angles by using overlapping ranges of $\sin^2\theta/\lambda^2$.

Interpolation between the calculated values of $\langle I \rangle$ is aided if they can be plotted on a straight line, which is approximately true if a Wilson plot is used. Interpolation between the points on the plot can be done quite satisfactorily by fitting a curve locally to three or four points. This is repeated for different sets of points along the plot. Extrapolation at low $\sin\theta/\lambda$ is helped by including the intercept of the least squares straight line through the points as an additional point through which the curve should pass.

For best results, it is essential that the interpolated values of $\langle I \rangle$ follow the actual calculated points even if these depart very much from a straight line. Special care must be taken in calculating E's at low angles. If these are systematically over-estimated this could easily result in failure to solve apparently simple structures. These E's are normally involved in more phase relationships than other reflexions and therefore have a big influence on phase determination. The number of strong E's chosen for phase determination is normally about 4 x no. of independent atoms + 100. More than this may be needed for triclinic or monoclinic crystals.

5. Setting up phase relationships.

Multan uses the very efficient algorithm for generating 3-phase structure invariants due to Dewar (1968). Each relationship will be set up three times unless care is taken to restrict the search to unique relationships only. Even so, the same relationship may be set up more than once when symmetry-related reflexions are present in the array. This gives the opportunity of applying the correct space group weights (Giacovazzo, 1974) to the relationships. These are given by

$$w_{hk} = (\varepsilon_{\bar{h}}\varepsilon_{k}\varepsilon_{h-k})^{-1/2}\sum_{p}\sum_{q}\delta_{pq}\exp[2\pi i(-\underline{h}^{T}\underline{d}_{p} + \underline{k}^{T}\underline{d}_{q})] \qquad (2)$$

where $\delta_{pq} = 1$ when $\underline{h}^{T}(I - C_{p}) = \underline{k}^{T}(I - \underline{C}_{q})$
$= 0$ otherwise

the summations are over all the space group symmetry operations which are represented by the point group operators C_p and translation vectors $\underline{d}_p$.

The correctly weighted relationships are obtained as the sum of the symmetry-related relationships set up by the algorithm just described (Main, to be published). Use of the correct weights leads to more accurate phases, although the effect diminishes with increasing size of structure.

6. Finding reflexions for phase determination.

It may happen that some reflexions enter into very few phase relationships. The phases of such reflexions are therefore poorly determined in terms of other phases and so they are eliminated at this point. This is done by the convergence procedure of Germain, Main & Woolfson (1970).

7. Assignment of starting phases.

Whether a small number of starting phases are used or all of the phases are included in the starting set, values are assigned to unknown phases using magic integers. The concept of magic integers was introduced by White & Woolfson (1975). For a sequence of n integers, $m_1, m_2 \ldots m_n$, n phases are represented in terms of a single variable, x, by

$$\phi_i = m_i x \mod(2\pi) \tag{3}$$

For any set of phases, ϕ_i, the equations are approximately satisfied by some value of x in the range $0 \leqslant x < 2\pi$. The nature and size of the errors involved has been investigated by Main (1977) who also gave a recipe for deriving magic integer sequences which minimise the rms errors in the represented phases.

To assign phase values, the variable x in equation (3) is given a series of values at equal intervals in the range $0 \leqslant x < 2\pi$. The enantiomorph may be defined by exploring only half of the n-dimensional phase space, ie. $0 \leqslant x < \pi$. For each value of x there corresponds a set of n phases ϕ_i. An efficient magic integer sequence ensures that these phase sets are as different from each other as possible.

For very large sets of phases, the magic integer sequence used is formed from the consecutive integers n to 2n-1, where n is the number of phases represented. These will all give an rms error of about 90°. This is little different from assigning random numbers as phase values, but it is easier to program and is always reproduceable.

8. Phase determination.

Phase determination is carried out using a weighted tangent formula:

$$\phi(\underline{h}) = \text{phase of} \left\{ \sum_k w_k w_{h-k} E(\underline{k}) E(\underline{h}-\underline{k}) \right\} \tag{4}$$

where w_h = weight associated with $\phi(\underline{h})$.

The variance of the determined phase $\phi(\underline{h})$ may be calculated according to Karle & Karle (1966) from

$$V(\underline{h}) = \frac{\pi^2}{3} + 4 \sum_{n=1}^{\infty} \frac{(-1)^n}{n^2} \frac{I_n(\alpha(\underline{h}))}{I_o(\alpha(\underline{h}))} \tag{5}$$

where $\alpha(\underline{h}) = 2\sigma_3\sigma_2^{-3/2}|E(\underline{h})||\sum_k w_k w_{h-k} E(\underline{k})E(\underline{h}-\underline{k})|$

The correct weight to use for the determined phase is inversely proportional to the variance and, to a good approximation, this is proportional to $\alpha(\underline{h})$. In addition, when the values of $\alpha(\underline{h})$ become large, the weights are further adjusted to make $\alpha(\underline{h})$ match the estimated values, α_{est}, calculated from the known disribution of 3-phase structure invariants (Cochran, 1955). The weights used, therefore, are given by

$$w_h = \min(0.2\alpha(\underline{h}),\ 1.0,\ (\alpha_{est}+5)/\alpha(\underline{h})) \qquad (6)$$

and this is shown in the Figure.

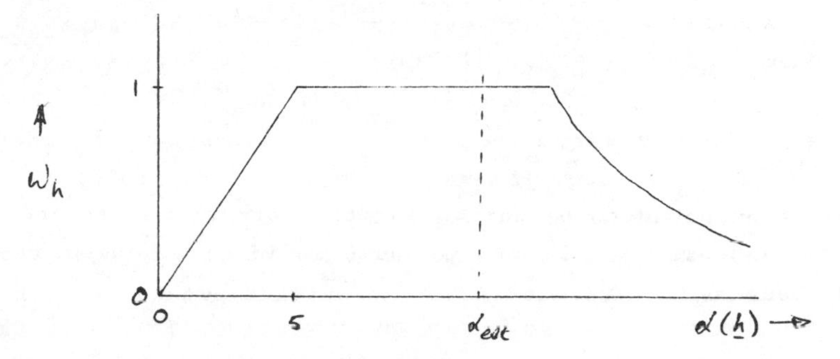

The constant of proportionality between w_h and α (0.2) is arbitrary, but the value given works well for most structures. As $\alpha(\underline{h})$ exceeds its expected value, the weight decreases. This has the effect of making $\alpha(\underline{h})$ agree more closely with its statistically expected value than it otherwise would. This generally results in more accurate phases. It is particularly valuable if the structure contains a heavy atom and it prevents all phases refining to zero in space groups like P1.

When all phases, instead of a small subset, are assigned initial values to start the phase determination, the complete set of phases effectively defines the origin and enantiomorph. It is therefore unnecessary to designate some phases specially for origin definition. The tangent formula (4) is used for phase refinement and all phases are given the same weight for the first cycle. This weight is calculated so it is 1.5 x rms wt. of phases at end of cycle. The weights at the end of the cycle are calculated from (6). Any phases with weights less than the rms value are kept fixed. On subsequent cycles, this threshold is gradually lowered so more phases are refined each time. Also, all weights are decreased at the end of each cycle unless they are calculated from (6) with a higher value. This decreases the effect of very inaccurate phases on the refinement of others and so increases the probability of obtaining the correct answer. This is a development of the Rantan procedure of Yao (1981).

9. Figures of merit.

Multan uses three figures of merit which are later combined into a single figure:

a)

$$\text{ABSFOM} = \frac{\sum_{h} \alpha(\underline{h}) - \alpha_r}{\sum_{h} \alpha_{est}(\underline{h}) - \alpha_r}$$

where $\alpha_r = \sum_{h} \sum_{k} \kappa^2_{hk}$

This is expected to be unity for the correct phases and zero for random phases. It is not a good discriminator between good and bad phase sets.

b)

$$\psi_0 = \frac{\sum_{h} \left| \sum_{k} E(\underline{k})E(\underline{h}-\underline{k}) \right|}{\sum_{h} \left(\sum_{k} |E(\underline{k})E(\underline{h}-\underline{k})|^2 \right)^{1/2}}$$

The summation over $\underline{k}$ includes the strong E's for which phases have been determined and the indices $\underline{h}$ are given by those reflexions for which $E(\underline{h})$ is very small. The numerator should therefore be small for the correct phases and will be much larger if the phases are systematically wrong. The denominator normalises ψ_0 to an expected value of unity.

c)

$$R_\alpha = \frac{\sum_{h} |\alpha(\underline{h}) - \alpha_{est}(\underline{h})|}{\sum_{h} \alpha_{est}(\underline{h})}$$

This is a residual between the actual and the estimated α's. The correct phases should make R_α small.

If a set of phases is found for which $\psi_0 <$ 1.25 and $R_\alpha <$

0.20, Multan assumes this is the correct answer and immediately calculates the map without completing the remaining phase sets. If it happens to be the wrong answer (very unusual), phase determination can be completed on a subsequent run. When phase determination is completed, the figures of merit are combined as

$$CFOM = w_1 \frac{ABSFOM - ABSFOM_{min}}{ABSFOM_{max} - ABSFOM_{min}} + w_2 \frac{\Psi_{max} - \Psi_0}{\Psi_{max} - \Psi_{min}} + w_3 \frac{R_{max} - R_\alpha}{R_{max} - R_{min}}$$

where the weights w_1, w_2, w_3 are normally set at 0.6, 1.2, 1.2 respectively. This gives values of CFOM in the range 0 to 3 with the best sets of phases having the higher values.

10. Interpretation of E-maps.

The complete interpretation of the E-maps is done in three stages:

a) peak search

b) separation of peaks into potentially bonded clusters

c) application of simple stereochemical criteria to identify possible molecular fragments

The stages a) and b) are very similar to those described by Declercq, Germain, Main & Woolfson (1973) and give a list of peak coordinates. Stage c) is described by Main & Hull (1978). Here, stereochemical criteria of maximum and minimum bond lengths and angles are applied to the E-map peaks to identify chemically sensible molecular fragments. Peaks which do not fulfil the criteria are eliminated in such a way as to maximise the number of high peaks which are accepted. Different peaks may be assumed to be spurious, giving rise to alternative interpretations. The interpretation and a plot of the peaks on the least squares plane are all output.

11. Least squares tangent formula.

If Ψ_0 and R_α are good figures of merit, then a process of phase refinement which minimises these quantities ought to produce good phases. A formula to achieve this was derived by the author and investigated by Stuart Fiske (1982) with promising results. Quite independently, Tate and Woolfson have produced a related formula and obtained even better results. The simplest form of this new tangent formula can be derived as follows.

Set up the system of equations

$$D(\underline{h}) = \sum_{k} E(\underline{k})E(\underline{h}-\underline{k}) \tag{7}$$

which is obviously related to Sayre's equation (Sayre, 1952). The magnitude of $D(\underline{h})$ is estimated statistically (related to estimated α) and its phase is $\phi(\underline{h})$. With correct phases, these equations should be satisfied. A phase refinement process can therefore be set up which will minimise

$$R = \sum_{h} |\varepsilon(\underline{h})|^2 \tag{8}$$

where $\varepsilon(\underline{h}) = D(\underline{h}) - \sum_{k} E(\underline{k})E(\underline{h}-\underline{k})$

A minimisation of R with respect to the phases leads directly to the formula (the least squares tangent formula)

$$\phi(\underline{h}) = \text{phase of} \left\{ |D(\underline{h})| \sum_{k} E(\underline{k})E(\underline{h}-\underline{k}) + |E(\underline{h})| \sum_{k} E(\underline{k})\varepsilon(\underline{h}-\underline{k}) \right\} \tag{9}$$

The first term is identical to the usual Karle-Hauptman tangent formula and the second term vanishes when the phases are exact. Not only large E's may be used in this formula, but equations (7) may be set up for which $E(\underline{h})$ is very small or zero (corresponding to ψ_0). They will not contribute to the main term in (9), but the $\varepsilon(\underline{h}-\underline{k})$ will contribute to the second term. This allows the active use of weak reflexions in phase refinement and it significantly increases the power of phase determination.

12. References.

Cochran, W. (1955) Acta Cryst., 8, 473-478.

Declercq, J-P., Germain, G., Main, P. & Woolfson, M.M. (1973) Acta Cryst., A29, 231-234.

Dewar, R.B.K. (1968) Thesis: University of Chicago.

Fan Hai-Fu, Yao Jia-Xing, Main, P. & Woolfson, M.M. (1983) Acta Cryst., A39, 566-569.

Fiske, S. (1982) Thesis: University of York.

Germain, G., Main, P. & Woolfson, M.M. (1970) Acta Cryst., B26, 274-285.

Germain, G. & Woolfson, M.M. (1968) Acta Cryst., B24, 91-96.
Giacovazzo, C. (1974) Acta Cryst., A30, 631-634.
Hull, S.E. & Irwin, M.J. (1978) Acta Cryst., A34, 863-870.
Karle, J. & Hauptman, H. (1956) Acta Cryst., 9, 635-651.
Karle, J. & Karle, I.L. (1966) Acta Cryst., 21, 849-859.
Koch, M.H.J. (1974) Acta Cryst., A30, 67-70.
Main, P. (1977) Acta Cryst., A33, 750-757.
Main, P. (1978) Acta Cryst., A34, 31-38.
Main, P. & Hull, S.E. (1978) Acta Cryst., A34, 353-361.
Sayre, D. (1952) Acta Cryst., 5, 60-65.
White, P.S. & Woolfson, M.M. (1975) Acta Cryst., A31, 53-56.
Yao Jia-Xing (1981) Acta Cryst., A37, 642-644.

DIRDIF: APPLICATION OF DIRECT METHODS TO DIFFERENCE STRUCTURES FOR THE SOLUTION OF HEAVY ATOM STRUCTURES, AND EXPANSION OF A MOLECULAR FRAGMENT

Paul T. Beurskens

Crystallography Laboratory, Toernooiveld
6525 ED Nijmegen, The Netherlands

Co-authors: W.P. Bosman, H.M. Doesburg, R.O. Gould, Th.E.M. van den Hark, P.A.J. Prick, J.H. Noordik, G. Beurskens, V. Parthasarathi, H.J. Bruins-Slot, R.C. Haltiwanger, M.K. Strumpel, and J.M.M. Smits

Abstract. When part of a structure is known, direct methods can be used to solve the unknown part of the structure. Often the known part of the structure consists of one or more heavy atoms, either on general, or on special or pseudo-special positions. The known part of the structure may also consist of a molecular fragment, found by ab-initio direct methods, or by Patterson rotation search techniques.
The difference structure factors, phased by the partial structure, are used as input to a weighted tangent-refinement procedure for phase extension and for the refinement of input phases and amplitudes. If the known atoms do not uniquely determine the structure, symbolic addition techniques are introduced to solve the (pseudo-symmetry) ambiguities.

The method is referred to as "DIRDIF". It is most useful if the known part is only marginaly sufficient to solve the structure, or if the known atoms lie in special or pseudo-special positions (origin ambiguity), or if, for noncentrosymmetric structures, the known atoms form a centrosymmetric arrangement (enantiomorph ambiguity). The automatic computer program uses observed structure amplitudes, and positional parameters of the known atoms to produce a greatly improved electron density map, which is passed on to a peak search and interpretation routine.

Vector search rotation functions, and translation functions in DIRDIF Fourier space, are incorporated in the program system.

1. Introduction. Applications of direct methods to difference structure factors require at least a reconsideration of all direct methods techniques. For instance, how to set up a sigma-2 listing, while the absolute values of the normalized structure factors change with each cycle of tangent refinement? Well-established procedures for, say, convergence mapping, origin and enantiomorph fixation, and selection of the starting set of phases, are not valid when applied to difference structure factors. E.g.: is the origin fixed, when heavy atoms are located at 0,0,0; $\frac{1}{3}$,0,0; $\frac{2}{3}$,0,0; in space group $P\bar{1}$? At present, most of these problems are solved, and the results are published and available in the computer program (see Beurskens et al., 1982, and references therein).

In this section, the principles underlying the DIRDIF method, are illustrated by a simple example, where the dominating heavy atoms are in special positions.

Example "NORA" $Au(S_2CN(C_4H_9)_2)_2\ Au(S_2C_2(CN)_2)_2$ (Noordik and Beurskens, 1971) Space group $P2_1/c$, Z=2. The reflections hkl (h=2n, k+l=2n) were very strong relative to the remaining reflections. The gold atoms were placed at the positions 000, ½00, 0½½ and ½½½. A conventional difference Fourier synthesis would have resulted in a fourfold superposition of the structure. A unique solution was obtained as follows.

For "strong" reflections, the local average $<|F_{obs}|>$ is plotted as a function of $\sin\Theta/\lambda$, see Figure 2.

This plot is used to bring the data on an absolute scale. The calculated structure factors for the given gold positions are:

$$F_{calc} = 4f_{Au} \exp(-B_{Au} \sin^2\Theta/\lambda^2).$$

where B_{Au} is the mean isotropic temperature factor of the gold atoms. For the set of "strong" reflections, both sign and magnitude of the light atom contribution are given by

$$F_{rest} = |F_{obs}| - F_{calc}(Au)$$

since in this case F_{calc} is always positive.

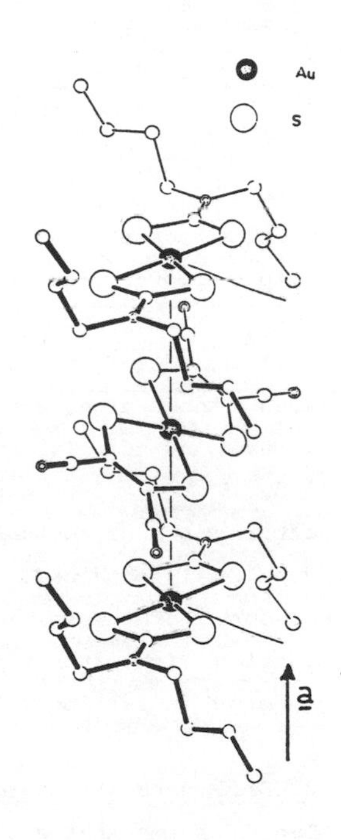

Figure 1. "NORA"

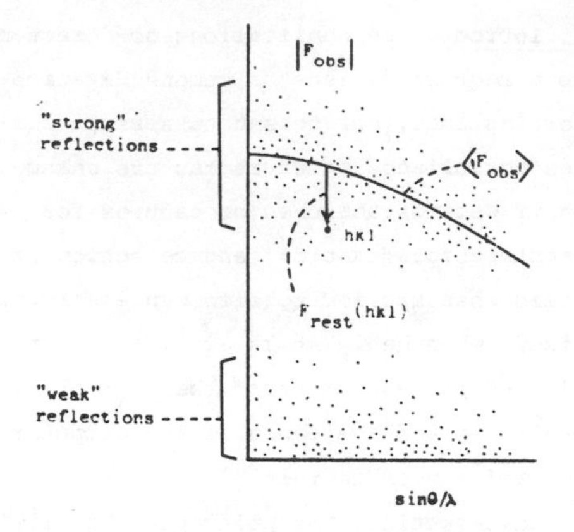

Figure 2 Graph of a number of $|F_{obs}|$ values. One example hkl is shown to have a negative F_{rest}.

For "weak" reflections, the contributions of the gold atoms to the structure factors cancel, $F_{calc} = 0$. The structure amplitude for the rest structure is given by:

$$|F_{rest}| = |F_{obs}|$$

A Wilson plot for these reflections gave the overall temperature factor of the light atoms (S, N, C, H). The $|F_{rest}|$ values were normalized, i.e. converted to $|E_{rest}|$ values. Where direct methods cannot be applied to the total structure, they can be applied to the rest structure. The sigma-2 or triple-product sign relationship was applied to all E_{rest} values with $|E_{rest}|$ greater than, say, 1.0. At this stage, the set of "strong" reflections contained 367 reflections with $|E_{rest}| > 1.3$, having a known sign, and all, of course, in parity classes eee and ooo (e=even, o=odd). Two reflections of the set of "weak" reflections (221 and 348, $|E_{rest}| > 2.0$) were arbitrarily given a sign to fix the origin. Thereafter, signs of another 180 reflections in the set of "weak" reflections were easily obtained.

A Fourier synthesis, referred to as DIRDIF Fourier synthesis, revealed the complete light atom structure. A generalisation of this "special" procedure is given in section 4.

2. Difference-structure factors; weights

Let us assume that a small part of a structure is known. The partial structure factor, F_p, is the structure factor calculated for the known atoms. $|F_{obs}|$ is the observed structure amplitude. In conventional Fourier and refinement techniques $|F_{obs}|$ is given the phase ϕ_p (phase of F_p). In general, the true phase of F_{obs} deviates from the calculated phase ϕ_p.

In Figure 3, the phase difference between the true F_{obs} and the calculated F_p is β. Assuming that there are no errors in the positions of the known atoms, the structure factor of the unknown part of the structure (i.e. the difference structure, or rest structure) is given by

$$F_r = F_{obs} - F_p \tag{1}$$

Of course, β is not known and one cannot calculate F_r. Let us consider the smallest and the largest possible value for F_r (see Figure 4):

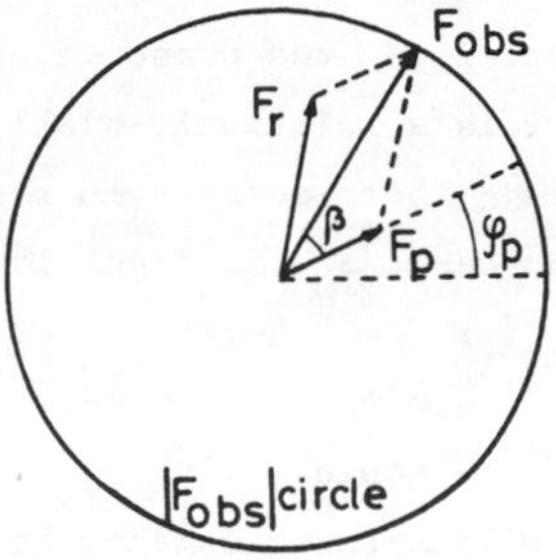

Figure 3. Vector diagram for Equation (1).

$$\Delta F_1 = |F_{obs}| \exp i\phi_p - F_p \tag{2a}$$

$$\Delta F_2 = |F_{obs}| \exp i(\phi_p + \pi) - F_p \tag{2b}$$

Note, that $|\Delta F_1| \leq |F_r| \leq |\Delta F_2|$.

Considering the phase of F_1, ϕ_1, we have to distinguish two sets of reflections.

Set 1 $|F_p| < |F_{obs}|$: $\phi_1 = \phi_p$ (3a)

Set 2 $|F_p| > |F_{obs}|$: $\phi_1 = \phi_p + \pi$ (3b)

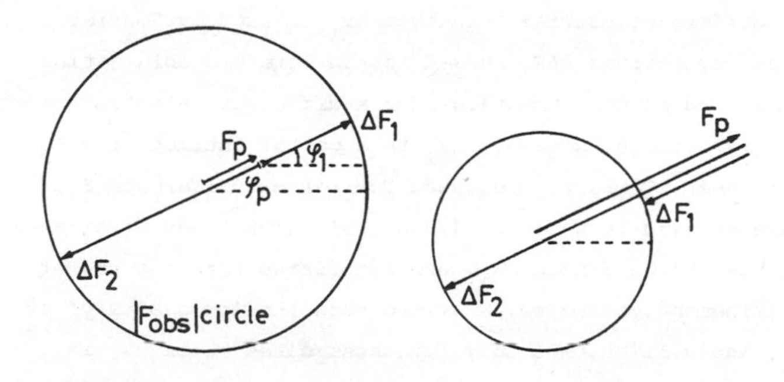

set 1: $|F_p| < |F_{obs}|$ set 2: $|F_p| > |F_{obs}|$

Figure 4. Vector diagrams for Equation (2).

When a small part of the structure is known, most of the reflections belong to set 1 and the phase of ΔF_1 equals the calculated phase ϕ_p.
Set 2 is a relatively small, but important set of reflections.
(Note: all zero-observed reflections belong to set 2; they play a role only in case $|F_p|$ is significantly greater than the standard deviation of $|F_{obs}|$.)

Before ΔF_1 can be calculated (Eq. 2), the data has to be brought on an absolute scale.
The Wilson-Parthasarathy formula (Parthasarathy, 1966) is used for a least-squares refinement (Gould et al., 1975) of
SC = the scale factor,
B_p = the overall temperature parameter of the known part of the structure,
B_r = the overall temperature parameter of the difference structure.
A careful evaluation of these quantities leads to reliable ΔF_1 values. After refinement of SC, B_p and B_r, and the calculation of ΔF_1, the data will be normalized.
Define, and calculate:

$$E_1 = \Delta F_1/g \tag{4}$$

where g is the usual normalization factor for the rest structure. It is the purpose of DIRDIF to find the true E_r value; i.e. the phased, normalized structure factor of the rest structure. After the refinement, the resulting E_r values are converted to F_r values:

$$F_r = g\, E_r \tag{5}$$

Weights

In the conventional difference Fourier technique ΔF_1 is used as Fourier coefficient. For many reflections, ΔF_1 indeed is the more probable estimate for F_r, and may be used in the summations for a difference electron density map. For other reflections, however, ΔF_2 is a better estimate for F_r, and ΔF_1 should not be used. Correctly weighted Fourier or difference-Fourier coefficients are derived by Woolfson (1956), Sim (1960) and Srinivasan (1968). Similar weights, to be denoted W_1, are associated with the use of ΔF_1 in the DIRDIF refinement procedure. W_1 represents the 'probability' of ΔF_1 relative to ΔF_2, and will be used only for those cases where ΔF_1 is more probable than ΔF_2.
The phases of the reflections of set 2 ($|F_p| > |F_{obs}|$) are very reliable, especially if $|\Delta F_1|$ is significantly greater than zero (say, $|E_1| > 0{,}7$): for these reflections we use $W_1 = 1.0$.
The weights W_1 are used in the initiation of the weighted tangent refinement.

3. Refinement and phase extension

All reflections with $|E_1|$ greater than, say, 0.9, are subjected to tangent refinement and phase extension. E_1 is used as input to the weighted tangent formula if $|E_1|$ and its weight W_1 are sufficiently large. New phases, denoted ϕ_r, are calculated for all reflections with $|E_1| > 0.9$. In general, ϕ_r is different from ϕ_1, and is a better estimate for the true ϕ_r value. Reflections with very low weight W_1 ($F_p = 0$) now may have a well defined phase (phase extension).

Any change in the phase of a reflection necessitates an adjustment of its amplitude: for each reflection the new ϕ_r value is used to calculate the corresponding $|E_r|$ value, see Figure 5. The new E_r values are subject to errors arising from incorrect E_1 values and from in-exact tangent-formula results. The refinement, however, is to be repeated. The 'reliability' of the tangent-formula result is denoted W_r.

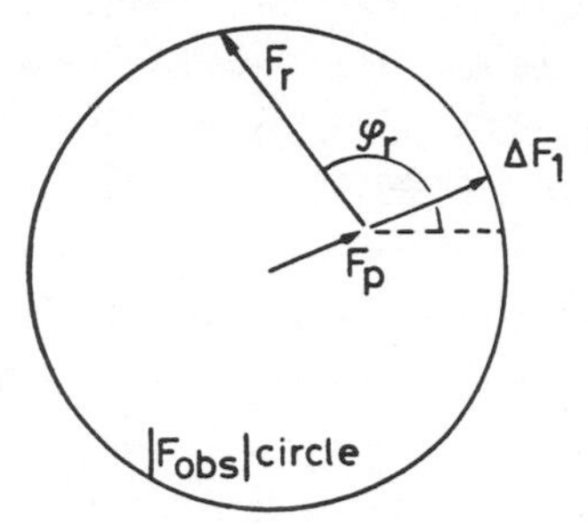

Figure 5. Construction of F_r for a given phase ϕ_r.

Whether or not the new phases ϕ_r are accepted depends on the corresponding weights. If $W_r > W_1$ then the new phase ϕ_r is accepted with weight W_r. If $W_r < W_1$ then the new ϕ_r value is only partially accepted, i.e. ϕ_r is replaced by a value inbetween ϕ_1 and ϕ_r (if permitted by space group symmetry).

The new ϕ_r, $|E_r|$ and W_r values (if $|E_r|$ and W_r are sufficiently large) are input to the second cycle of the tangent refinement, to obtain renewed ϕ_r values, and so on. Convergence usually is achieved in four or five cycles. As a consequence of the definitions of W_1 and W_r, the original phases ϕ_1 still play an important role in the second and following cycles, and phases of reflections of set 2 cannot change much.

Presumably, this is the reason that the tangent refinement applied to difference structure factors, appears to be very stable in regard to the usual enantiomorph problems encountered in space groups such as P1 and $p2_1$.

The final E_r values are the 'most probable E_r values':

they are transformed back to F_r values (Eq. 5), and the corresponding Fourier synthesis will give the 'most probable electron density map' for the rest-structure.

A typical application of DIRDIF is given in the following example.

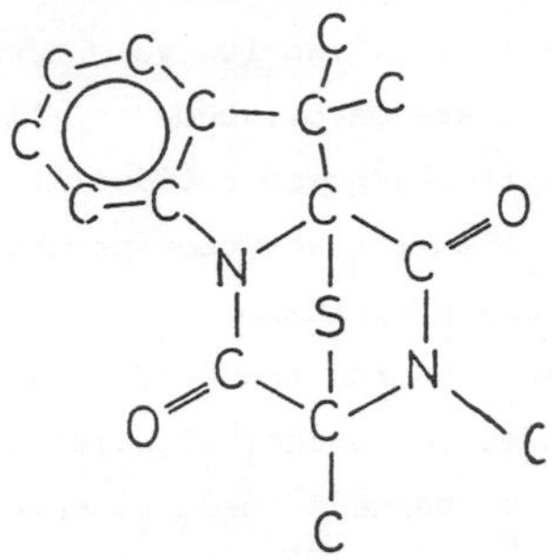

Example "MONOS" $C_{15}H_{16}N_2O_2S$

(Noordik et al., 1978)

Space group $P2_12_12_1$ Z = 4.

This structure could not 'routinely' be solved by direct methods. The position of the sulphur atom was found from a Patterson map as: x ∿ 0.0, y = 0.09, z = 0.14. This position, however, is a pseudo-special position as the trial model includes a mirror plane at x = 0 (and a center of symmetry at ¼, ¼, 0). A conventional difference Fourier would have led to a superposition of the structure with its enantiomorph. This mirror symmetry was destroyed by shifting the atom about 0.15 Å (x_{new} = 0.02); by this choice (x positive rather than negative) the enantiomorph is fixed.

The application of DIRDIF by an automatic run of the computer program gave the following results:

- Partial structure factors were calculated for all 1684 reflections, including the unobserved (= zero-observed) reflections.
- Refinement of SC, B_p and B_r gave: B_p = 4.81 $Å^2$ for the sulphur atom, and B_r = 4.17 $Å^2$ for the remaining atoms.

(Note: we found in other structures that unreasonable values for B_p and B_r result from errors in cell contents, or errors in the trial structure, which thereby may be detected).

- The data were normalized; 331 reflections had $|E_1| > 0.9$ and were subjected to the tangent refinement procedure.
- The final 331 E_r values were transformed to F_r values; for the remaining reflections Sim - Srinivasan coefficients were calculated; and a Fourier synthesis was calculated.
- The top 20 peaks correspond to the 19 atoms of the molecule (including the sulphur atom) and one false position.

4. The trial structure forms a sub-cell:

Origin fixation.

In inorganic and coordination chemistry one often finds one or more heavy

atoms at special or pseudo-special positions, such that these atoms form a sub-cell of the unit cell.
Structure factor calculation shows that the trial structure does not contribute to all parity groups. A conventional difference Fourier synthesis would lead to a superposition of the structure with the translated structure.
Only the heavy atoms show the extra translational symmetry; the rest structure has to be positioned relative to the known heavy atoms. Origin fixation in DIRDIF means the specification of an origin-ambiguity. The general principles are extremely simple: from the subset of reflections with no contribution from the trial structure, the strongest ten reflections are given a letter symbol to represent their phases. These reflections are input to the tangent formula, together with the reflections having reliable E_1 values.
At the end of the first cycle the relationships between the letter symbols, arising from the application of the symbolic addition technique, are analysed. Because of the origin ambiguity, there are always two or more equally probable solutions. The origin is fixed by arbitrarily choosing one of these solutions, and the letter symbols are eliminated in terms of phases. The second and following cycles are performed with numerical phases, as described in section 3.

5. A centrosymmetric trial structure for a non-centrosymmetric crystal: enantiomorph discrimination

Let us consider the following, rather common, situation. In space group $P2_1$ we have located one heavy atom. A conventional difference Fourier synthesis would lead to a superposition of the structure with its enantiomorph. DIRDIF with enantiomorph discrimination (Prick, Beurskens and Gould, 1983) is a very convenient tool for solving this problem.
Assume that the center of symmetry of the trial structure coincides with the origin. (If not, the origin will be shifted during the execution of the program.) Then all calculated structure factors F_p have phases $\phi_p = 0^\circ$ or 180°.
In order to destroy the center of symmetry we must find reflections with phases, significantly different from 0° or 180°. Such reflections are recognized by:

a. a rather low weight W_1 (i.e. a relatively small $|F_p|$ value), and
b. a rather inconsistent set of sigma-2 interactions (i.e. the tangent formula cannot determine whether ϕ_r is 0° or 180°).

We select some ten reflections and assign letter symbols to represent their phases; these reflections are input to the tangent formula, together with the reflections having reliable E_1 values. At the end of the first cycle relationships between the letter symbols will be found, and they are interpreted such that the numerical values of the letter symbols are 90° or -90°. The enantiomorph is selected by choosing one of the two equally probable solutions for the letter symbols. Thereafter the numerical phases will be refined.

6. Tangent formula recycling by DIRDIF

Application of direct methods for the solution of not too small structures often leads to an electron density map (or E map) from which a molecular fragment can be recognized, and tangent formula recycling techniques (Karle, 1968) were introduced to complete the solution of the structure.
The application of DIRDIF to light-atom structures is equivalent to tangent-formula recycling for difference-structure factors. There are important differences between the conventional tangent recycling and the present procedure: different reflections and sigma-2 interactions are used, the number of unknown atoms that must be found is less, and the refinement is stable with respect to origin and enantiomorph.

Direct methods problems often are caused by regular patterns in the structure, enhanced by planarity or symmetry of the molecules: a molecular fragment is then easily recognized, and by "subtraction of the fragment from the structure", the cause of troubles is at least partly removed. DIRDIF, therefor, is a powerful alternative to tangent recycling.

When a molecular fragment cannot be expanded one may assume that either the space group is wrong or that the fragment is misplaced. In either case it is useful to expand the reflection data to a half sphere, and to execute the methods described above in space group P1. Automated translation functions, based on the DIRDIF-P1 Fourier map, can be used to position the known fragment (Doesburg and Beurskens, 1983).

7. Automated vector search rotation functions

The availability of crystallographic data bases, and programs for prediction of molecular geometries, are causing a revival of Patterson methods. The Nordman & Schilling (1970) vector search programs have been automated (Strumpel et al.,1983) and incorporated in the DIRDIF system. To our experience, the vector search rotation functions are very reliable, and certainly less computer time demanding than generally believed.

The molecular fragment of known geometry is oriented to give maximum

FLOW DIAGRAM OF THE DIRDIF PROGRAM SYSTEM

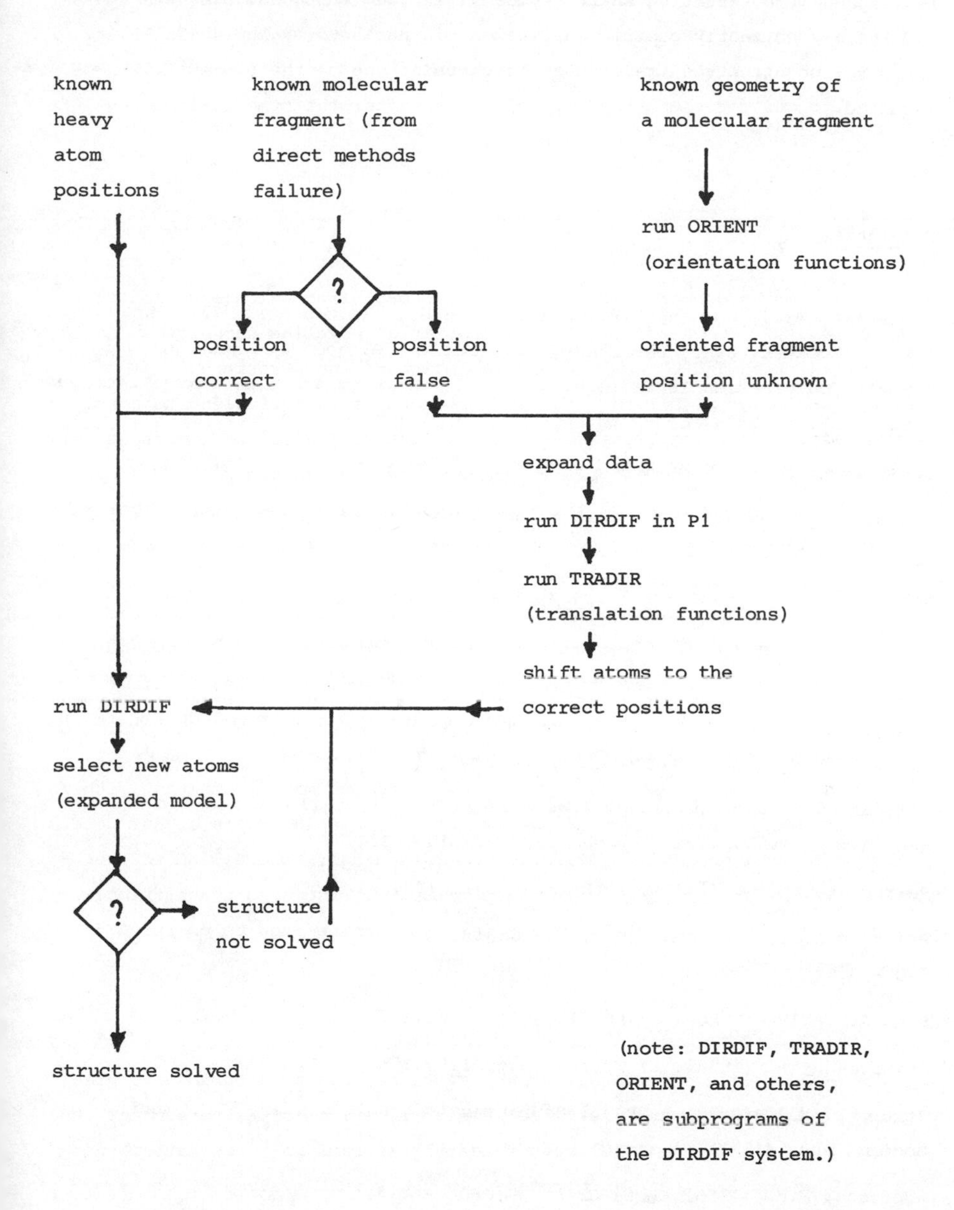

consistency with the Patterson function. The resulting set of atomic parameters is used for the calculation of translation functions in DIRDIF Fourier space; the resulting shift vector is applied to the atomic parameters, and the now correctly positioned fragment is expanded by the same DIRDIF phase refinement procedure. Manual intervention, although possible, is not required.

References

-Beurskens, P.T., Bosman, W.P., Doesburg, H.M., Van den Hark, Th.E.M.,Prick, P.A.J., Noordik, J.H., Beurskens, G., Gould, R.O. and Parthasarathi, V. (1982). Conformation in Biology. Eds. R. Srinivasan and R.H. Sarma. Adenine Press, N.Y. p. 389-406.

-Doesburg, H.M. and Beurskens, P.T. (1983). Acta Cryst. A39, 368-376.

-Gould, R.O., Van den Hark, Th.E.M. and Beurskens, P.T. (1975). Acta Cryst. A31, 813 - 817.

-Karle, J. (1968). Acta Cryst. B24, 182 - 186.

-Noordik, J.H. and Beurskens, P.T. (1971). J. Cryst. Mol. Struct. 1, 339 - 345.

-Noordik, J.H., Beurskens, P.T., Ottenheijm, H.C.J., Herscheid, J.D.M. and Tijhuis, M.W. (1978). Cryst. Struct. Comm. 7, 669 - 677.

-Nordman, C.E. and Schilling, J.W. (1970). In: Crystallographic computing (Ed. Ahmed) Munksgaard, Copenhagen, p. 110 - 114.

-Parthasarathy, S. (1966). Z. Kristallogr. 123, 27 - 50.

-Prick, P.A.J., Beurskens, P.T. and Gould, R.O. (1983). Acta Cryst. A39, 570 - 576.

-Sim, G.A. (1960). Acta Cryst. 13, 511 - 512.

-Srinivasan, R. (1968). Z. Kristallogr. 126, 175 - 181.

-Strumpel, M., Beurskens, P.T., Beurskens, G., Haltiwanger, R.C., and Bosman, W.P. (1983). Abstracts, E.C.M.8, Liège.

-Woolfson, M.M. (1956). Acta Cryst. 9, 804 - 810.

DIRECT METHODS AND SUPERSTRUCTURES

R. Böhme

Institut für Mineralogie der Ruhr - Universität
Universitätsstr. 150, D - 4630 Bochum, FRG.

Introduction

Using direct methods we look at the structure factors as complex numbers of given magnitudes and unknown phases and estimate the values of sums of suitable phases. These estimations are more reliable when the magnitudes are large as when they are small - large or small relative to the value they maximal can have. Given the structure factor equation for the F's

$$F(\underline{h}) = \sum_{j=1}^{n} f_j \cdot \exp(2\pi \underline{h} \cdot \underline{x}_j)$$

we know that the complex structure factor $F(\underline{h})$ is the sum of the contributions of the n atoms in the unit cell. So it is the sum of n complex numbers, what we can illustrate in a vector diagram (fig.1). When we say the phase estimations are reliable when the magnitudes of the reflexions involved are large, so it means in our diagram that the phases of all the n contributions have to be near to that of the final structure factor and in case the estimations are less reliable and the magnitudes of the reflexions involved are small, the phases of the different contributions vary quite a lot from the final phase value of the sum.

The structures showing superstructure effects are those where we have a more or less large amount of reflexions which are weak but where it is important that these reflexions are not handled as of zero - intensity. When we do so, we always get a structure model that shows some kind of higher symmetry. Different approaches

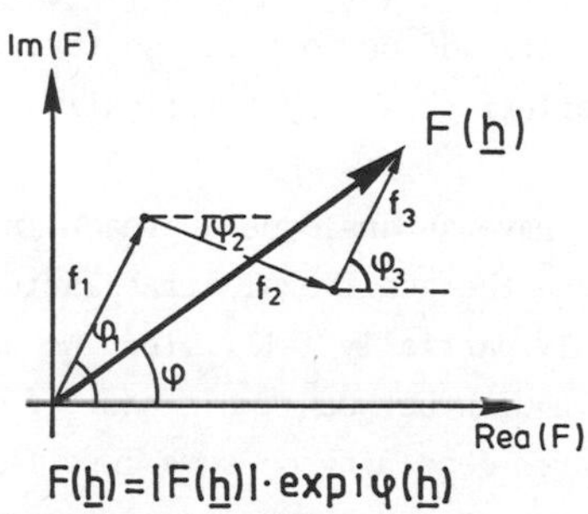

fig.1: vector diagram of F

have been made in the past th overwhelm this problem also by using direct methods. If the superstructure effects arise because e.g. a heavy atom in a special position simulates an additional translation, the contribution of that atom can be subtracted from the measured structure factors and the phase estimation can be carried out for the difference structure provided the position of the heavy atom is well known (Beurskens & Noordik 1971, Beurskens 1984). We suggest here a desciption from a symmetrical point of view. It explains a lot of anomalous effects that may occur during a structure determination procedure. The procedure suggested solves structures using direct methods without any knowledge of the atoms showing higher symmetry. The method presented is therefore also practicable, when the strong reflexions determine the positions of the atoms of the structure only approximately or when some of the equivalent atoms of ideal symmetry have to be eliminated. The often successful method of rescaling the E - values (Hauptman & Karle 1959) is a special case of our procedure under conditions which can easily be verified.

Description depending on symmetry

We get systematically extincted reflexions when we choose a centered cell for indexing the reflexions. That can also be shown in a vector diagram (fig.2). If the motif - the molecule, part of the structure or so - here indicated by a triangle - is repeated four times in the unit cell by 1/4 of a lattice vector, then the reflexions h of our onedimensional example are strong when h equals 4n, because the contributions of all motifs to the structure factors have the same directions. And when h does not equal 4n, the contibutions of the four motifs add up to zero, so that the reflexions are systematically absent.

We have an analogue approach in the case the additional translation is only partially fullfilled. We distinguish between two types of structures depending on some important features of the pseudotranslations producing the superstructure effects.

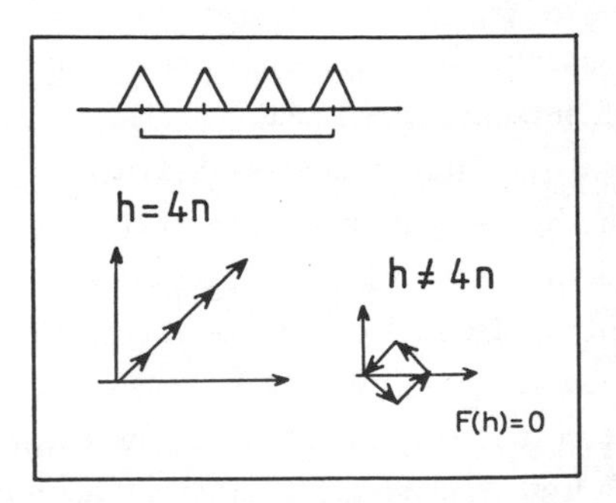

fig. 2

Type 1

A fraction of the atoms in the unit cell satisfies an additional pseudotranslation $\underline{t}$ or the atoms satisfy $\underline{t}$ approximately (Böhme 1982). A two-dimensional example showing a pseudotranslation $\underline{t}$ = (o,1/4) is given in fig. 3. Because all electron density that satisfies the pseudotranslation exactly does not contribute to the weak reflexions, we distinguish only between two classes of reflexions, the set of the "strong" or "main" reflexions $\underline{H}$ and the set of the "weak" or "superstructure" reflexions $\underline{U}$.

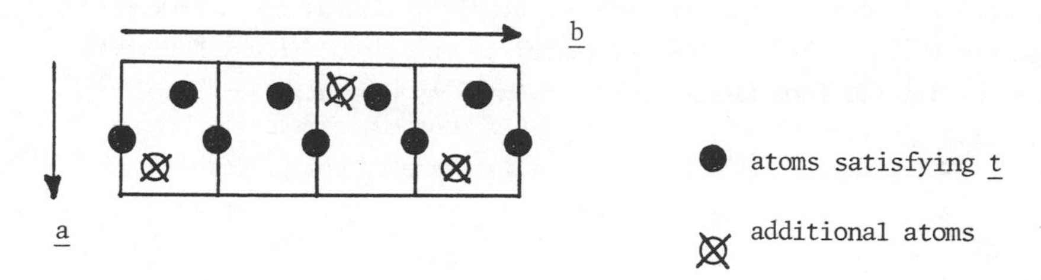

fig. 3: twodimensional structure showing a pseudotranslation $\underline{t}$ = (0,1/4)

The two classes of reflexions are normalized separately. That is equivalent to estimating the mean intensity by

$$a \cdot \exp\left(-B_o \sin^2 \vartheta / \lambda^2\right) \cdot \sum_{j=1}^{n} f_j^2$$

instead of only $\sum_{j=1}^{n} f_j^2$ (when there are n atoms of scattering power f_j in the whole unit cell). Estimating the mean intensity - separately for main and superstructure reflexions - is necessary when no information about the atoms satisfying the pseudotranslation is available a priori. That is particularly the case when the pseudotranslation is fullfilled only approximately. The "effective" scattering power depends then on how far the atoms are away from the ideal positions. It can be shown by graphical integration e.g. that the maxima are still of elliptical shape in any case the pseudotranslation is of index 2. An explicit example is given in fig. 4.

The set of all triplets can be divided into three classes according to the number of main and superstructure reflexion they contain:

class (a) three main reflexions
(b) three superstructure reflexions
(c) two superstructure and one main reflexions

In case the index of the pseudotranslation is 2 , triplets of type (b) do not exist because of parity conditions. Under such circumstances the set of triplets does not determine the phases of the superstructure reflexions completely. The phases of all the weak reflexions may hve to be shifted by π . Otherwise the atoms may be shifted away from the ideal position in a direction that is just opposite to the correct one. It can be shown that also the triplets of type (c) are approximately well estimated via Sayre's equation (Hai-Fu et al. 1983), although they contain structure factors (or E'values) of different functional forms. Estimations via probabalistic formulas are given recently by Gramlich (1984) and Giacovazzo (1984).

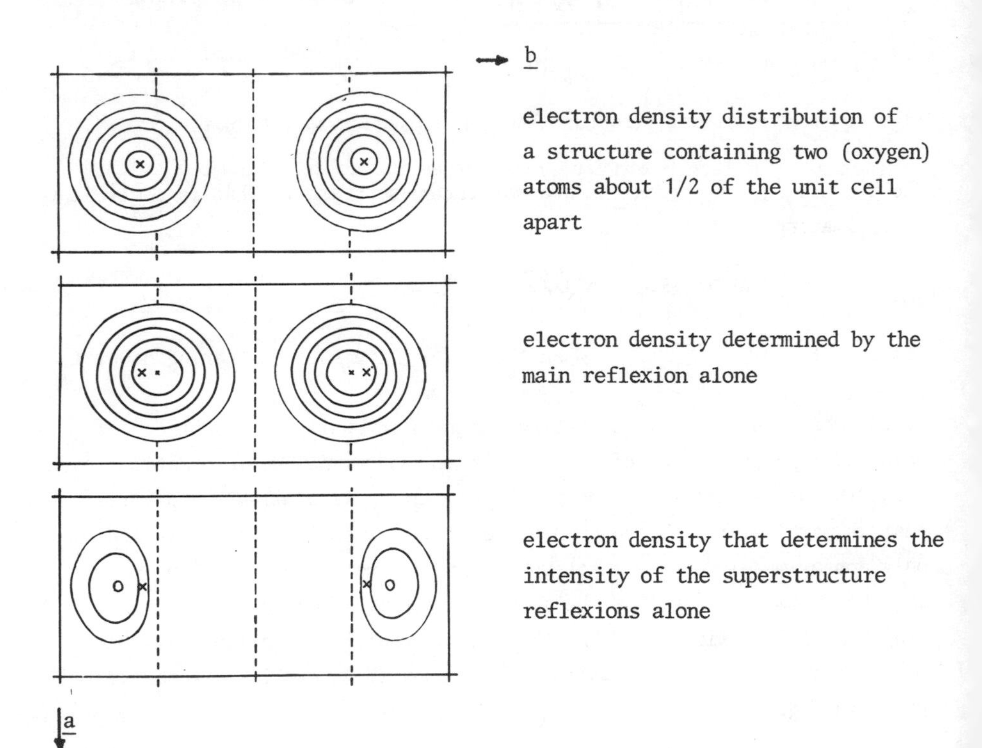

b
electron density distribution of a structure containing two (oxygen) atoms about 1/2 of the unit cell apart
electron density determined by the main reflexion alone
electron density that determines the intensity of the superstructure reflexions alone
a

fig. 4: Two dimensional example showing a pseudotranslation $\underline{t}$ = (0,1/2). x = atomic positions; atoms assumed in $\underline{x}_1$ and $\underline{x}_2$, so that $\underline{x}_1 \approx \underline{x}_2 + \underline{t} + 0.2$ Å .

Type 2

The pseudotranslation t is of index p and not all p atoms equivalent by t are available (Böhme 1983). A hypothetical twodimensional example showing a pseudotranslation t = (o,1/6) is given in fig. 5. In this case p classes

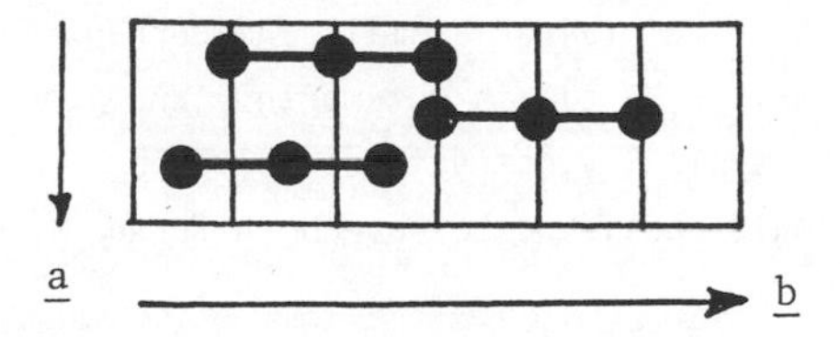

fig. 5

of reflexions of different intensity are to be handled (renormalized or normalized separately). The triplets are estimated well via the normal formulas. But each reflexion has to be provided with a phase shift that is typical for the class the reflexion belong to. Otherwise a part of the structure or molecule may appear in the correct orientation but in a wrong position in the unit cell.

A direct methods procedure that contains all descripted features has been presented (Böhme 1984).

Literature

P. Beurskens & J. Noordik (1971) Acta Cryst. A27, 187

P. Beurskens (1984) DIRDIF article

R. Böhme (1982) Acta Cryst. A38,318

R. Böhme (1983) Z. Naturforsch. 38a, 304

R. Böhme (1984) Acta Cryst. A40, C423

F. Hai-Fu, Y. Jia-Xing, P. Main & M.M. Woolfson (1983) Acta Cryst. A39,566

H. Hauptman & J. Karle (1959) Acta Cryst. 12, 846

V. Gramlich (1984) Acta Cryst. A40 in press

C. Giacovazzo (1984) in preparation

INTRODUCTION TO PATTERSON SEARCH METHODS

By Christer E. Nordman

Department of Chemistry, University of Michigan, Ann Arbor, Michigan 48109
U.S.A.

Introduction

In present day crystallography, dominated by highly successful direct-methods packages, Patterson methods have maintained a small and perhaps diminishing niche. That this niche exists at all is noteworthy, and its contents deserve some study, because, typically, structures solved by Patterson methods appear to be difficult ones, and to have been found untractable to direct methods.

The basis of the power of Patterson methods is that they allow *a priori* structural information to be brought to bear on the structure-solving process more effectively than do direct methods. If one or more rigid molecular fragments are known, their calculated vector patterns can be used as probes in exhaustive searches of the Patterson function, stored in the computer memory. Similarly known, local non-crystallographic symmetry can be searched for in Patterson space.

The requirements, then, of a computer implemented Patterson interpretation are a Patterson function, a suitable expression of the known feature to be searched for, and a criterion for judging the goodness of fit. In this paper we consider these three elements, and the ways in which they are, or might be, formulated, as well as some newer developments in Patterson search methods. Two recent reviews (Nordman, 1980a; Nordman & Hsu, 1982) provide additional material on the subject.

The treatment here is almost entirely limited to methods which operate in Patterson, or vector space. Search methods operating in reciprocal space, such as the rotation function (Rossmann & Blow, 1962) and modifications thereof, widely used in protein crystallography, are not extensively covered.

The Patterson function: Preparation, storage and retrieval

To preserve maximum detail and minimize distortion near the origin, the Patterson function should be sharpened, and the origin peak removed. Coefficients of $E^2(hk\ell) - 1$, optionally damped by $\exp(-D\sin^2\theta/\lambda^2)$, with a damping parameter $D \approx 2 - 4$ Å^2, are convenient in small-molecule applications. A macromolecular Patterson is best computed by subtracting local averages from the sharpened F^2-values. Such averages may be taken over local ranges of $\sin\theta/\lambda$ or local regions of $hk\ell$, the latter representing anisotropic treatment. Thus, the Patterson coefficients are taken as $[F^2_{sh}(hk\ell) - \langle F^2_{sh}\rangle_{hk\ell}]\exp(-D\sin^2\theta/\lambda^2)$ where the $\langle F^2_{sh}\rangle_{hk\ell}$ are averages of the sharpened F^2-values in local regions of $hk\ell$.

Whether any other a priori modification of the Patterson function or its coefficients is likely to benefit the searches is an open question.

One asymmetric unit of the Patterson function is typically stored in the computer memory. As a guide to choosing the fractional cell-edge grid on which the Patterson function is evaluated, we consider the resolution of an individual Patterson peak.

An isolated vector peak in a fully sharpened or slightly damped Patterson function at Cu sphere (0.8 Å) resolution has a width at half-height of about 0.7 Å. At three-quarter-height the width is 0.45 Å (Nordman & Schilling, 1970). To insure that the peak is present in the computed Patterson map at 75 percent of its height, the grid spacing must not exceed $0.45/\sqrt{3}$ Å, assuming a uniform orthogonal grid. As a general rule, no grid spacing should exceed $d/3$, where $d = (2\sin\theta_{max}/\lambda)^{-1}$ is the resolution of the data. Spacings of about $d/4$ are recommended; even denser grids should be used, if interpolation is not employed in retrieving the function values.

The stored Patterson map of a large high-resolution structure can contain on the order of 10^5 function values; protein Pattersons even more. Storing this array is done most economically by packing several map values into one memory word. On machines with a 32-bit word (e.g., IBM, VAX), individually addressable 8-bit bytes are particularly convenient. The stored values (0,...,255) preserve the experimental accuracy and allow meaningful expression of gradient components, difference Pattersons, etc. On 36-bit machines, 6-bit units (0,...,63) have been used with simple shifting and masking operations doing the packing and unpacking required on storage and retrieval. For greatest economy and speed, without interpolation, 2-bit (0,...,3), or 3-bit units (0,...,7) have been successfully used (Braun, Hornstra & Leenhouts, 1969; Hornstra (1970); Egert, 1983).

The retrieval of Patterson function values by interpolation in the stored map is a compromise between speed and accuracy. Interpolation schemes using 1 (no interpolation), 4, 8, 11 and 27 map points have been used (Nordman, 1980b). The 8-point interpolation is the simplest interpolation routine which returns a continuous function, a feature which is important in vector-space refinement.

Information Input and Image-Seeking Functions

Methods for computer analysis of the Patterson function by means of superposition maps have been developed by Jacobson (Jacobson & Beckman, 1979; Jacobson & Richardson, 1984) and by Simonov (1982) and co-workers. Methods for handling superpositions on multiple (as distinct from single atom-atom) vectors, and multisolution algorithms have yielded solutions of noncentric structures with $\sim$60, centric with $\sim$160 non-hydrogen atoms per cell. While a particularly heavy atom is not required, the procedure benefits from the presence of unequal atoms. No structural information is used.

When the chemistry of the high-resolution structure under study is known to the extent that a rigid molecular fragment can be assumed, the full potential of the Patterson search method can be exploited. Sources of structural information are the Cambridge Data File, or known bond lengths and angles, augmented, when applicable, by empirical force field calculations (Egert, 1983).

In selecting a search fragment it is well to remember that fragments located on the outer extremities of a molecule tend to have high thermal motion, and give rise to relatively weak vector sets. The higher the vector density of the set, including vector overlaps, the more easily will it be recognized in the Patterson function. Planar or rod-like (e.g., α-helices) fragments have high-density vector patterns and are usually easier to find than bulkier fragments with the same number of atoms.

For an N-atom search fragment, there are $N(N-1)/2$ intra-fragment vectors. An approximate profile for each vector peak can be computed from a knowledge of the atomic numbers of the two atoms, the data resolution, and the estimated thermal parameter (Nordman & Schilling, 1970). These profiles are used to compute the amount of mutual overlap between any two vector peaks whose centers are within the radius of the overlapping peak. In this way the calculated density at each vector peak center is augmented.

We now have a set of calculated vector densities, or weights, w, at the $N(N-1)/2$ vector points. Many of these vector points lie so near one another that they are unresolved, and if used in a search, would effectively sample the same point in the Patterson. The vector set is therefore thinned out by placing the vectors in descending order of w and deleting from the list all vectors below a given one whose distance from the given vector is less than a specified minimum $\approx$ d or somewhat less. From the resulting list, the lowest-weight vectors may be eliminated. In rotation searches it is often advisable to remove very short vectors, as these are not very discriminating, and tend to fall on distorted regions in the Patterson. The result, then, is a selected set of well distributed vector components and weights.

This vector calculation procedure must be modified for macromolecules, partly because the computing task of evaluating all vector pair overlaps goes up very sharply with N. If atom coordinates are known for the macromolecular search fragment, the latter can be placed in a large cell and its structure factors computed. The Patterson is computed by Fourier synthesis, and search vectors selected as above.

The number of interatomic vectors in the unit cell is proportional to N^2, if N is the number of atoms in the cell. Since the volume of the cell is approximately proportional to N, it follows that the number of vectors per unit volume of the Patterson function increases as N. It also follows that if a search fragment comprises one tenth of the structure, its vector set comprises one hundredth of the Patterson. The problem of optimizing the probability of detecting a very weak fragment vector set in a very noisy environment then becomes a central one in Patterson search.

Many "image seeking functions" (ISF), or measures of fit of a search vector set w(i), i = 1,...,n, to the Patterson function P, have been used in different laboratories. Several have been discussed in the above-cited reviews.

Two are of particular interest. One is the sum of products

$$\sum_{i=1}^{n} w(i)P(i) \qquad (1)$$

where P(i) is the (interpolated) Patterson value at the point of the vector i of weight w(i). (1) is closely related to $\sum_{i=1}^{n} P(i)_{model}P(i) \approx \int P_{model}P dV$ which is the correlation function employed in the reciprocal space rotation function and other reciprocal-space methods.

The other is

$$MIN(m,n) = \sum_{i=1}^{m} P(i)/\sum_{i=1}^{m} w(i) \qquad (2)$$

where $m \leq n$ and the sums include those in terms for which the ratio P(i)/w(i) takes on its m lowest values. The lower this ratio, the worse is the accommodation of the model vector w(i) to the Patterson. By basing the ISF on the subset of the m worst-fitting search fragment vectors the selectivity is enhanced. Clearly, this requires sorting the n ratios at each step in the search, at some additional computing cost.

While the ISF (2), typically with $m \approx (0.1 - 0.5)n$, has proven effective, no claim can be made that it is the optimal form of an ISF.

In an attempt to assess the performance of (2) and related ISF-expressions, a series of simulated Patterson searches using random number sequences has been carried out (Nordman, 1983). Let the simulated Patterson function be represented by $P(i) = S(i) + N(i)$ where $S(i)$ and $N(i)$ are sequences of 500 random numbers with positive means $\langle S \rangle$ and $\langle N \rangle$ and standard deviations $\sigma(S)$ and $\sigma(N)$. The $n = 300$ simulated search vectors $w(j)$ are a contiguous (without loss of generality) subset of $S(i)$, such that $w(j) = S(j+k_0)$, i.e., the $w(j)$ sequence is a part of the $S(i)$ sequence starting at $k_0 + 1$. The 300 values $w(j)$ are taken as the "known" search vector weights. The search is carried out by translating the $w(j)$ sequence along the $P(i)$ sequence, allowing the $w(j)$ to sample the $P(j+k)$ for successive values of k, from 0 to 200. At each of the 201 steps, the data are sorted, and several ISF's evaluated. A given ISF is judged successful if it assumes a higher value for $k = k_0$ than for any of the 200 other values of k. The percentage of success achieved by a given ISF in several hundred independent searches allows us to compare different ISF's with each other. Among the ISF's evaluated were (2), a "weighted" version

$$WMIN(m,n) = \sum_{i=1}^{m} [w(i)P(i)]/\sum_{i=1}^{m} [w(i)]^2, \quad m \leq n \tag{3}$$

and a "minimum" version of (1):

$$MSP(m,n) = \sum_{i=1}^{m} [w(i)P(i)] \quad m \leq n \tag{4}$$

In the calculations $N(i)$ was Gaussian $\langle N \rangle = 300$ and $\sigma(N) = 100$. $S(i)$, including the vector weights $w(j)$ was also Gaussian with $\langle S \rangle = 3\sigma(s)$, and $\sigma(S)$ ranging from 15 to 50.

The results are given in Fig. 1. It is interesting to note that all curves have their maxima at $m < n$. For MIN, the optimal choice of m is approximately 0.3 n, in good agreement with experience from actual Patterson searches. A comparison of the results for MIN and WMIN suggests that WMIN is slightly more discriminating, but the difference is of marginal significance, at best.

At $m = 300$, the $MSP(m,n)$ function is identical to the ordinary sum of products of eq.(1). The fact that $MSP(n,n)$ is consistently poorer than any of the other ISF's tested tends to support the conclusion that Patterson-space methods have an advantage over reciprocal space methods in that they permit the use of more discriminating measures of fit.

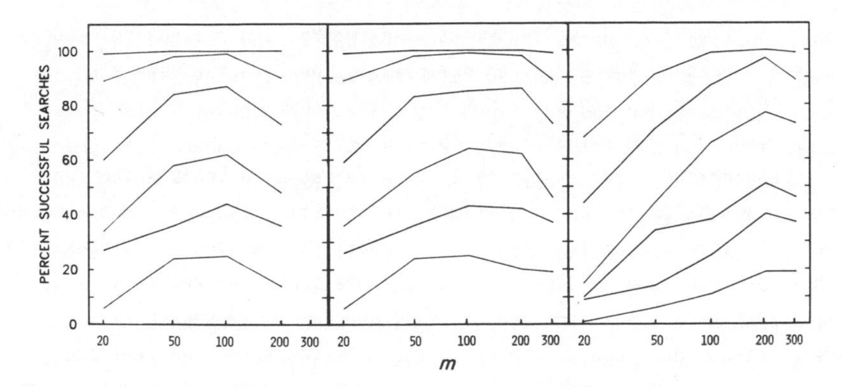

Fig. 1. Percentage of successful searches of random number sequences, as a function of $m \leq n = 300$ for three ISF's as defined in the text: MIN (left), WMIN (center) and MSP (right). For each ISF the success percentage is shown for $\sigma(S)/\sigma(N) = 0.15$ (bottom graph), 0.20, 0.25, 0.30, 0.40, and 0.50 (top graph). Each graph represents a minimum for 100 runs.

Search Procedures and Innovations

Placing a rigid molecular fragment in the unit cell involves 6 parameters, 3 rotational and 3 translational. For the purpose of a Patterson search the 6-dimensional problem can be separated into a 3-dimensional rotation search and a 3- (or lower) dimensional translation search. This can be seen from the fact that the orientation of the fragment can be determined from an orientation search with the intra-fragment vectors only, the latter being known without reference to the translational coordinates of the fragment.

The orientation is usually expressed in terms of Euler angles; unfortunately there are many different conventions in use, as there are for the definition of the Cartesian coordinate system in relation to the crystal axes. Taking the University of Michigan vector search programs as an example, we define the Cartesian axes X, Y, Z as X|| $\underset{\sim}{a}$, Y in the $\underset{\sim}{a}\,\underset{\sim}{b}$ plane (Y|| $\underset{\sim}{b}$ if $\gamma = 90°$), and Z || ($\underset{\sim}{a}$x$\underset{\sim}{b}$). The Euler rotations A, B, C are, in order:

A is a rotation about the X axis

B is a rotation about the rotated Y axis

C is a rotation about the rotated X axis.

A and B are the longitude and latitude of the direction of the rotated X axis on the surface of a sphere, with B = 0° and 180° at the poles. The range of B is 0-180°, of A and C 0-360°, in general.

If the search vector set has an n-fold rotation axis, and this axis is directed along X, then it can be seen that the necessary search range in C is reduced to 0-360°/n. Such an axis is present whenever the molecular fragment has an n-fold rotation or rotatory inversion axis, including $\bar{2} \equiv m$. For example, if the fragment is planar and laid out with its normal ($\bar{2}$) along

X, the C-range will be halved. Symmetry in the Patterson function reduces the search ranges of A and B. For example, if the Patterson symmetry is

2/m $0 \leq A < 360°$ $0 \leq B \leq 90°$

mmm $0 \leq A \leq 180°$ $0 < B \leq 90°$.

A translation search is typically carried out with vectors from one oriented fragment to another, related to the first by a space group symmetry element. If this symmetry element is a center of inversion, the translation search is 3-dimensional, if it is an axis (rotation or screw) it is 2-dimensional, and if it is a mirror or glide plane, the search is 1-dimensional. For example, in space group $P2_12_12_1$ three 2-dimensional translation searches would be made; these would each yield two of the three translational parameters, and thus allow an internal consistency check of the placement of the fragment.

With two symmetrically independent oriented fragments, chemically like or unlike, the translation search with vectors between them is 3-dimensional. An interesting case arises when 3 (or more) fragments are present. Now the translation vectors from fragment 1 to 2, 2 to 3 and 3 to 1 must form a closed triangle, a condition which serves as an internal consistency check on the positioning of all three.

Some recent maodifications of the traditional procedure are the following.

At the orientation search stage it is sometimes advantageous to modify a Patterson function by removing from it certain vectors. This is accomplished by modifying the coefficients $E^2(hk\ell)-1$ so as to remove the contribution of a specified list of vectors, and their symmetry equivalents, from the coefficients. An example is the structure of the tetramethoxy-stilbene-TCNQ complex, in which two search fragments are planar and parallel, as well as having similar vector patterns. It was shown (Strumpel, 1983; Strumpel, Zobel & Ruban, 1980) that removal of the oriented vector set of one fragment greatly improved the rotation search for the other.

An innovation of potentially wide application was introduced by Admiraal & Vos (1983). A 40-nonhydrogen atom tetrapeptide structure was solved by using two linked peptide units as a search fragment. As this is not a rigid fragment, the two angles ϕ and ψ at the central Cα were systematically varied in 10° steps, giving a total of 190 conformations tried. This somewhat tedious procedure has recently been automated (Egert & Sheldrick, 1984).

The end result of the orientation and translation searches described in the preceding paragraphs is a partial structure composed of the oriented symmetry-related search fragments properly placed in the unit cell.

Whereas it is entirely possible to proceed by Patterson methods, finding additional atoms by multiple superposition, it is generally simpler to proceed by direct methods, or, where feasible, by Fourier synthesis methods.

Recently, Patterson orientation search routines have been incorporated into direct methods packages. A highly automated orientation search routine has been built into the DIRDIF program system (Strumpel, Beurskens, Beurskens, Haltiwanger & Bosman, 1983). Also, a Patterson orientation search routine has been interfaced with the SHELX System (Egert & Sheldrick, 1984). In both of these cases the translational parameters of the search fragment (or fragments) are found by direct methods.

References

Admiraal, G. & Vos, A. (1984). Acta Cryst. C39, 82-87.

Braun, B. P., Hornstra, J. & Leenhouts, J. I. (1969). Philips Res. Rep. 24, 85-118

Egert, E. (1983). Acta Cryst. A39, 936-940.

Egert, E. & Sheldrick, G. M. Submitted to Acta Cryst.

Hornstra, J. (1970). Crystallographic Computing, edited by F. R. Ahmed, pp. 103-109. Copenhagen: Munksgaard.

Jacobson, R. A. & Beckman, D. E. (1979). Acta Cryst. A35, 339-340.

Jacobson, R. A. & Richardson, J. W. (1984). Am. Cryst. Assn. Abst. Ser. 2, Vol. 12, p. 40.

Nordman, C. E. (1980a). Computing in Crystallography, edited by R. Diamond, S. Ramaseshan & K. Venkatesan, pp. 5.01-13. Bangalore: Indian Academy of Sciences.

Nordman, C. E. (1980b). Acta Cryst. A36, 747-754.

Nordman, C. E. (1983). Proc. Indian Acad. Sci. (Chem. Sci.), 92, 329-334.

Nordman, C. E. & Hsu, L.-Y.R. (1982). Computational Crystallography, edited by D. Sayre, pp. 141-149. New York: Oxford University Press.

Nordman, C. E. & Schilling, J. W. (1970). Crystallographic Computing, edited by F. R. Ahmed, pp. 110-114. Copenhagen: Munksgaard.

Simonov, V. I. (1982). Computational Crystallography, edited by D. Sayre, pp. 150-158. New York: Oxford University Press.

Strumpel, M. K. (1983). Dissertation, Free University of Berlin.

Strumpel, M. K., Beurskens, P. T., Beurskens, G., Haltiwanger, R. C. & Bosman, W. P. (1983). Am. Cryst. Assn. Abst. Ser. 2, Vol. 11, p. 51.

Strumpel, M. K., Zobel, D. & Ruban, G. (1980). Am. Cryst. Assn. Abst. Ser. 2, Vol. 8, p. 37.

TRANSLATION SEARCH BY INTEGRATED PATTERSON AND DIRECT METHODS

By Ernst Egert,

Institut für Anorganische Chemie der Universität, Tammannstrasse 4,
D-3400 Göttingen, Federal Republic of Germany

Introduction

Patterson search has been shown by various authors to be a powerful tool for solving difficult crystal structures; its great strength is that it employs chemical information directly, and so can compensate for mediocre precision and resolution of the X-ray data (Egert, 1983, and references cited therein). Nevertheless, it is not nearly as popular as direct methods, which owe part of their success to automation and superior computational efficiency. We have attempted to combine the merits of both methods - in a manner that is generally applicable, efficient, automatic, computer-independent and easy to use - and thus to exploit all the *a priori* available information in order to solve large problem structures.

Description of the Search Procedure

In procedures to position a fragment of known geometry in the unit cell, the translation search has usually proved to be less reliable than the rotation search. Restricting ourselves to Patterson space, this is because the "cross"-vectors used to locate a fragment with respect to the origin suffer from errors in both the model geometry and orientation amplified by the symmetry elements; in addition, model vectors with very high weight are less likely than in the rotation search. In order to obtain atomic positions accurate enough for the subsequent structure expansion or refinement, either a fine search grid or some optimisation of promising solutions is necessary. Thus, if a Patterson sum or minimum function is used, the time for the translation search rises rapidly with the complexity of the structure. Furthermore, time-consuming interpolation procedures can hardly be avoided.

The calculated phases, in contrast to the way in which the Patterson function must normally be stored, are a continuous function of the atomic coordinates. When a fragment is moved through the unit cell keeping its orientation fixed:

$$F_H = F_H^o \cdot \exp 2\pi i H \cdot \Delta r$$

since all atomic displacements Δr are the same. So the structure factor F_H for any position is readily obtained by multiplication of F_H^o (the structure factor for the starting position) with a simple phase factor. For the true structure, the individual phases of the strongest reflections are linked by various statistical phase relations; amongst these, the three-phase structure invariants have proved to be especially useful:

$$\phi_H + \phi_K + \phi_{-H-K} \approx 0$$

The search fragment is usually incomplete and may also be not very accurate. Nevertheless, if its scattering power is significant, the triple phase relations should hold at least approximately for the correct solution, in the sense that the distribution of the phase sums is far from being random.

These considerations led us to the development of a novel strategy for Patterson translation search which, as far as we know, for the first time fully exploits in an integrated fashion the information contained in the sharpened Patterson function, the three-phase structure invariants and the allowed intermolecular distances (Fig. 1). This method is described in detail elsewhere (Egert & Sheldrick, 1984). In short, the position of an oriented search fragment is found by maximising the weighted sum of the cosines of a small number (40-60) of strong translation-sensitive triple phase invariants, starting from random positions. It is our experience that the triple phase refinement is well able to home in on fragments starting about 0.5 Å away from their true positions. This means that at least one trial per cubic Ångstrom is necessary in order to have a good chance of locating one search model correctly. A Patterson minimum function based upon intermolecular vectors is calculated only for those solutions that do not give rise to intermolecular contacts shorter than a preset minimum. This procedure avoids the time-consuming refinement in Patterson space and should be especially efficient for large structures. Finally the best solutions are sorted according to a figure of merit based on the agreement with the Patterson function, the triple phase consistency and an R-index involving E_{obs} and E_{calc}.

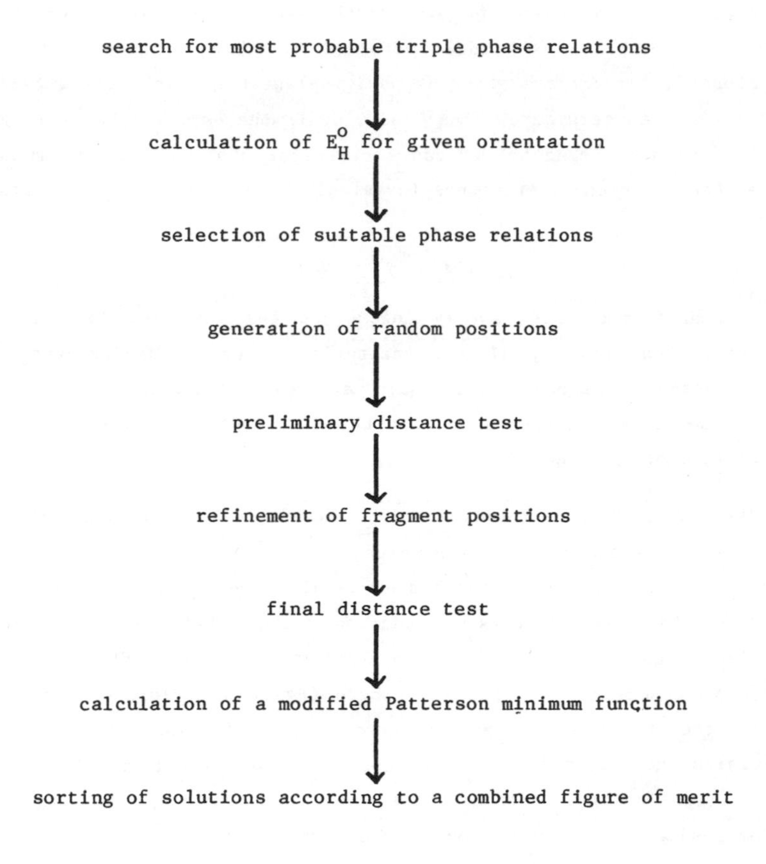

Fig. 1. Flow diagram of the translation search procedure

The procedure described differs from other Patterson translation functions (Nordman, 1966, 1980; Hornstra, 1970; Doesburg & Beurskens, 1983) in that the oriented model is placed with respect to all symmetry elements of the space group simultaneously. Tests with about 30 known structures, using search fragments taken from published crystal structures or from force-field calculations, have indicated that this routine is reliable and widely applicable. In particular, it is able to locate very large fragments (of more than 300 atoms) as well as single atoms even when the latter are not very heavy (e.g. P or S in large organic structures). In terms of computing times, it is also competitive with direct methods; under favourable circumstances it can even prove more economical. Above all, the variety of different criteria employed to judge solutions should make this novel combination of Patterson and direct methods a powerful structure solving strategy, if chemical information is available.

Features of the Program PATSEE

The procedure outlined has been combined with a conventional (Nordman, 1966; Hornstra, 1970) but highly automated real-space Patterson rotation search and implemented as a computer program called PATSEE which is valid and efficient for all space groups in all settings. Since it is written in a simple subset of FORTRAN it may be run without significant alteration on a wide range of computers, provided that sufficient memory (at least 40K words) may be addressed directly and that the word length is at least 32 bits. The program has been designed to be fully automatic, but the default settings may easily be changed by experienced users. The rotation search can find the orientation of a fragment of any size and allows one torsional degree of freedom. The translation search may locate up to two independent search models of any size (including single atoms), taking into account known atoms at fixed positions, if any. Since the program is compatible with SHELX-84, convenient facilities exist for the generation of the Patterson function before, and the structure expansion after, the fragment search. When the current extensive tests have been completed, PATSEE will be distributed together with SHELX-84.

I thank the Fonds der Chemischen Industrie for a Liebig-Stipendium and the Deutsche Forschungsgemeinschaft for financial support.

References

DOESBURG, H.M. & BEURSKENS, P.T. (1983). Acta Cryst. A39, 368-376.

EGERT, E. (1983). Acta Cryst. A39, 936-940.

EGERT, E. & SHELDRICK, G.M. (1984). Submitted for publication.

HORNSTRA, J. (1970). Crystallographic Computing, edited by F.R. Ahmed, pp. 103-109. Copenhagen: Munksgaard.

NORDMAN, C.E. (1966). Trans. Am. Crystallogr. Assoc. 2, 29-38.

NORDMAN, C.E. (1980). Computing in Crystallography, edited by R. Diamond, S. Ramaseshan & K. Venkatesan, pp. 5.01-5.13. Bangalore: The Indian Academy of Sciences.

LEAST SQUARES REFINEMENT

David Watkin
Chemical Crystallography Laboratory,
9 Parks Road, Oxford, ENGLAND.

Introduction

If a least squares refinement strategy were well defined, it would already have been programmed and become part of our scientific heritage. Instead, occasions continue to arise where a new strategy must be devised, and which can only (sometimes) be fully justified with hindsight.

This paper is intended to explain the role and application of least-squares (LS) procedures in crystallography. The mathematical coverage will deliberately be kept superficial, since most users must try to believe that the programs they use are theoretically sound, and there are many excellent books and articles detailing the mathematics. Here it is my ambition to outline the techniques, and to suggest how they may be effectively applied.

Aims of Refinement

There are perhaps two broadly different approaches to scientific research. In one, a hypothesis is made and justification is then sought experimentally. In the other, an explanation of some observable phenomenon is sought in terms of an intellectually acceptable model. The first approach explains why we do structure analysis, and the second, how.

In crystal structure analysis, we are attempting to provide a model which will explain the observed structure amplitudes. In addition, we can sometimes say with confidence that the subject of our current investigation should have features in common with a larger class, and we can treat these features as non-diffraction data which must be satisfied simultaneously with the diffraction data. Inevitably, the thresholds of acceptability will change from observer to observer, and from time to time. However, norms exist for the models; contemplating the deviations from these norms increases our understanding of the physical problem.

Structure analysis consists of two overlapping stages; structure solution and structure refinement. Structure solution consists essentially of identifying parameters needed to specify the model, and assigning them approximate values. Structure refinement involves 'improving' the parameter values, probably introducing additional parameters, and possibly eliminating others. Common usage has come to identify refinement with least squares, since that is the technique used in the final stages of the refinement of most high resolution and many protein structures. However, it is important to remember that the LS best solution is only one sort of best, and that there may be other solutions better suited to answer other questions posed of the model.

The steps in arriving at a suitable model are :

1. Deciding what the observations are, and what errors are associated with them.

2. Postulating a model appropriate to the question being posed, which will enable the observations to be simulated.

3. Deciding if the differences between the actual and the simulated observations are acceptably small. If they are, then we understand the phenomenon, otherwise we must try to modify the model.

Step 1 poses the problem of resolving the 'observation' from the 'model'. In X-ray crystallography, are the crystals 'Ideally Imperfect'? If an absorption correction is applied, is the correction part of the observation, or of the model? The answer to these questions really depends on what additional questions the model must answer, and upon the allocation of errors. For example, the errors associated with the Lp correction are generally regarded as being insignificant, so that the correction can be treated as an accurate, high precision, model designed to satisfy our observations of the diffraction process generally. The effect of the absorption correction is less well defined, and the errors associated with the 'observations' passed on to later stages should be adjusted to reflect this uncertain modification of the observed data. In particular, we should note that the errors in our data are now almost certain to be highly correlated.

Step 2, structure solution, is the subject of other papers.

Step 3 may require the use of several techniques, of which LS is only one. The final decision about the acceptability of the deviations between the observed and simulated observations is a personal one.

Factors Influencing Refinement.

The refinement process cannot be made to yield more information than that present in the data (including non-diffraction observations). Defficiencies in the data may include:

Systematic Errors

These really are shortcomings in the model. If the model being refined contains parameters highly correlated with the systematic errors, then the model will be prejudiced in a way depending upon the form of the systematic error. For example, failure to correct the diffraction data for the theta dependent component of the absorption effect will systematically reduce the atomic temperature factors. The LS technique is sensitive to systematic errors, and yields inaccurate results if they are present. A weighting scheme approximating to the full weight matrix may reduce the effect.

Random Errors

These are inevitable in any experimental observation. They should be minimised by careful design of the experiment (e.g. choosing appropriate crystal size, counting times and radiation). Often reducing random errors is something that must be purchased (by increasing the time spent in performing the experiment), or is in conflict with reducing systematic errors (large crystals giving good counting statistics may introduce non-random errors). LS should give an accurate, but not necessarily precise, result.

Shortage of Data

Generally, the LS technique, unlike Fourier methods, is insensitive to the absence of individual observations. Provided that proper weights are chosen for the distribution of errors in the observations and assumptions in the model, then the result is that the random errors in the data lead to random errors in the model. Increasing the amount of data, or its quality, will improve the model. Note that reducing the number of parameters in the model implies increasing the amount of data, since those parameters not refined will be given (either explicitly or implicitly) values that are themselves observations of high precision. Their accuracy will depend upon the approximations made in simplifying the model.

Lack of Resolution

If some of the data are not observable in a systematic way, this may mean that they will not adequately define some parameters in the model. The most common example is the link between the angular range of the diffraction data and the detail that can be resolved in the structure. Lack of resolution in the data usually shows itself as large esd's in the refined parameters, though these can also arise from a very inadequate model. For example, an anisotropic temperature factor may fail to give a satisfactory representaton for an atom that is disordered over two widely separated sites.

Least Squares

If we have n independent observations of a function in n parameters, then we can determine values for the parameters, and this is true even if the observations contain serious errors. Unless we have some pre-conceived ideas about the range of possible values for the parameters, we have no way of assessing their validity. In general, therefore, we try to have as large an excess of observations over unknowns as is practicable, and attempt to make our model best reflect these observations. Comparing the difference between the observed (Yobs) and simulated (Ycalc) values with the expected errors in the observations enables us to get a measure of the precision of the model, and also hopefully to ensure that the model is physically valid. For example, while three non-colinear points in 3-space will fit both a plane and a sphere, four might (but not necessarily) resolve the ambiguity. The LS solution for the model is that which minimises sigma $(Yobs-Ycalc)^2$. If absolute or relative errors can be associated with Yobs, each term should be appropriately weighted. If Y (any observable property) can be expressed as a linear function of x(j) (the parameters which describe our model), then, for the i'th of m observations we have a(i,1)x(1) + ... a(i,n)x(n) = Ycalc(i) ~ Yobs(i), (1)

where the matrix 'A(m,n)' of the independant variables is called the design matrix, and the observations 'Yobs' occur only on the rhs. Note that there is no assumption about the units of 'Y', and that we can in fact include observations of quite different physical properties, provided that we can express them as a linear function of the parameters 'x', and appropriately weight them. If 'Y' is a non- linear function of x, then the problem can be linearised through a Taylor expansion with neglect of high order terms:

a(i,1)x(1) + ... a(i,n)x(n) = (Yobs(i)-Yold(i)) (2)

Now the terms A(i,j) are the first derivative of Y(i) with respect to x(j), the right hand side is (Yobs - Yold) where Yold is the value of Y for a trial model, and the parameters x(j) are corrections to be applied to the model. Note that there are now terms due to the model on both sides, and that the weights must also reflect shortcomings in the model.

If Yold is not Ybest, the values of x(j) will not usually be the best shifts, so the procedure must be cycled. The process is not necessarily convergent, or may converge to a local minimum near to the trial model.

The most reliable way to solve 'm' simultaneous linear equations is via singular value decompositions, which use the full A(m,n) matrix. These methods are impractical in crystallography where both m and n are very large. For medium size structures, Newton's method, which involves inverting the normal (Hessian) matrix, H(n,n) = A'A, is generally used.

$$A'A.x = A'.Y, \qquad H.x = A'.Y, \qquad x = H^{-1}.A'.Y$$

Note that in forming 'H' some information is lost, and the inverse may be ill-determined or non-existent. Note also that since the derivatives are computed from the current model, the process depends critically on that model. For larger structures, 'H' is rearranged so that large elements lie close together along the leading diagonal, generally by the user selecting which parameters must be refined together.

Near zero elements are then eliminated and H partitioned into a series of smaller matrices, the 'block diagonal approximation'. In this approximation, Yold and all the derivatives still have to be computed. Having eliminated some elements, time is saved by forming fewer sums of products, and each block may be inverted independently. Futher time may be saved by only recomputing the vector A'Y on alternate cycles (since the terms in the matrix do not change much as convergence approaches), and perhaps also updating the diagonal elements of H. Approximations not using the full Hessian are subject to two major dangers. The most serious is that subdivision of the matrix may conceal a singularity or near singularity inherent in the formulation of the problem, for example unrecognised higher symmetry, or atoms lying in special positions (though these should be detected by a preprocessor). The second is that the additional approximations made in neglecting some off-diagonal terms, or only updating the matrix, make convergence less reliable. If the full matrix has been replaced by a large number of small sub-matrices, merely inverting this block diagonal matrix will be neither safe nor efficient, and better convergence will be achieved using conjugate gradient methods. Some sort of 'cascade' refinement may be used for even larger structures. In these procedures, some parameters are not refined every cycle, so that while all Yold have to be computed, only the derivatives of the refined parameters need be. Different groups of parameters are refined in different cycles, and if the groups contain some parameters in common, this method can be made to approximate to a full matrix method with modified matrix. An extension which further saves time is to store the A and B parts of the structure factors, which can then be updated rather than recomputed totally each cycle.

For the largest structures, all off-diagonal terms of H computed from the diffraction data may be neglected, and now conjugate gradient methods must be used. Time saved by not accumulating the full matrix is offset by the need to recompute Y and the derivatives, but Fast Fourier techniques may do this to sufficient accuracy. Off-diagonal terms due to restraining information may still be computed and used, though there is a problem in scaling them to the diffraction data. Because these elements have special relationships between them (through the physical model) they can be very conveniently stored using sparse matrix techniques. Structures may even be refined, as a regularisation procedure, against the restraints alone, though unless there are restraints involving symmetry operators or atoms in symmetry generated fragments, the matrix will be singular because the coordinate system origin is not defined.

A Procedure

1. Maximise the number of good observations.
2. Ensure you have an appropriate starting model.
3. Minimise the number of variables.
4. Use independent criteria to judge the validity of the refinement.

1. Although structures have been refined against the 'unobserved' data, it is much better to have good data. The diffraction data can be supplemented by other information.

a) Restraining information. Yobs in (2) is some non-diffraction function of the model parameters, usually a molecular parameter such as planarity, bond lengths and angles, or a function of the atomic temperature factors. Yobs can be obtained from the literature i.e. given a numeric value, or can be computed as a function of other model parameters (e.g. average bond length), and thus be used to encourage non-crystallographic symmetry on the model. The restraining equation (2) must be weighted to reflect the confidence in the assumption. Since restraining information usually links several parameters explicitly, it contributes to the off-diagonal elements of the normal matrix, and of course leads to large correlations between the parameters. The esd's of derived parameters must be computed from the full variance - covariance matrix. The restraint that Ycalc ~ Yold for a single parameter effectively limits the shift on that parameter by contributing to a diagonal element only, and is an elegant way of stabilising an oscillating or divergent refinement. Restraining information in conflict with the other data will have a large $\Delta(Y)$ after refinement, and so may be reconsidered.

b) Non-refined parameters. Not all the parameters needed to compute Yold in (2) need be refined. The non-refined value ascribed to a parameter then plays the role of an observation of very high weight. Because the parameter is not being refined, we have no simple indication about the appropriateness of the ascribed value. However, since non-linear LS converges to local minima, it is sometimes appropriate to fix some parameters to 'reasonable' values while others are being adjusted (e.g. fix U[iso] during initial refinement of X's, fix X's during initial refinement of U[ij]'s). Note that cascade and blocked matrix refinement

methods, by suppressing the correlation information, make implicit use of non-refined parameters and thus have less certain convergence properties. Parameters that are highly correlated (e.g. U and occupancy, pseudo symmetric atoms, scale and extinction) MUST be refined in the same matrix block to get esd's showing the worth of the parameter values. If the refinement of these parameters is unstable, shift limiting restraints may be imposed, or better, if the number of correlated parameters is not too large, they may be put in a single matrix block and refined by eigenvalue filtering. This approximates to singular value decomposition, but restricts the sum of the squares of the shifts (whereas svd minimises it), yielding a solution not too far from the model. The problems of nearly singular matrices can be reduced by suitable numerical techniques (Choleski, Marquardt, filtering), but are best resolved by a deeper awareness of the difficulty.

c) Making assumptions about the model, for example stating that some atoms are isotropic, or have coupled vibrations, or that the geometry is fixed. These are often called (hard) constraints.

2. Because there may be local minima in the minimisation function for non-linear LS, refinement may converge to an invalid result if the model is too far from the global minimum. Futher, since the current model modifies the rhs of (2), the better the model the more reliable the calculated shifts. The best model available should be used to initialise the LS, and this may be obtained by:

a) Doing Fourier or tangent refinement to get good atomic positions.

b) Regularising the structure either against external geometries, or to raise the internal symmetry.

c) Geometrically placing missing atoms. Usually only done for hydrogen, but much more widely applicable.

d) Making careful guesses at the temperature factors.

3. Though there are often simple economic grounds for restricting the number of parameters refined, there are sometimes also good scientific reasons. It is not necessarily true that a data set will define adequately a full parameterisation of the model, in which case the user may wish to simplify (U[iso] rather than U[ij]) or group (rigid body, TLS)

the parameters. 'Riding' models (in which the shifts for several different physical parameters are assigned to a single LS parameter), often used to refine hydrogen atoms, can be used for other parameters. For example, the temperature factors of corresponding atoms in two molecules related by a pseudo centre may be linked together, while the positions are refined individually (in the same block of course!). For some types of structure, it may be sensible to represent the temperature factors by an overall T tensor, and have separate L and S tensors for tightly linked sub-groups. Such a parameterisation would get round the difficulties in extracting T,L and S from U's, and serve to compute excellent bond length corrections. A valid reduction in the number of variables in a problem should increase the reliability of the remainder. Thus, even if hydrogen positions cannot be found and refined with the diffraction data, sensible positions based on chemical evidence should be used in forming Ycalc to improve the accuracy of the remaining variables. The fear that constraints may conceal real structural features may be well founded, but should be offset by the prospect of free refinement leading to worthless parameter values.

4. A low Hamilton 'R' for a structure is only one indication that the structure may be correct. Other tests must also be applied. The molecular parameters must be acceptable, and if not (assuming appropriate numerical methods were used), restraints or constraints used to see if traditional values are compatible with the X-ray data. If they are not, perhaps you have made a discovery, or have just got bad data.

Weights

Equation (2) may be weighted by pre-multipying both sides by W(m,m), the 'weight matrix'. This is usually difficult to obtain, and is often replaced by a diagonal matrix, which assumes uncorrelated errors in each of the m equations. Unit or statistical weights are used during model developement, so that $\Delta(Y)$ can be compared with $\sigma(Y)$. Once the parameterisation has been chosen, weights giving a constant value for $\langle w.\Delta Y^2 \rangle$, grouped in any rational way, are needed to give parameter error estimates FOR THAT MODEL. Other weighting schemes can be devised, for example, to increase the resilliance and range of convergence (iterative reweighting), or to enhance particular features (charge distribution studies).

Conclusion

There is of course no conclusion, except to be vigilant and sceptical. Modern programs tend to execute to completion, often taking care of problems unforseen by the user. Even so, it is the users responsibility to supervise the computation. The following table, based on several months detailed statistics, contains its own warnings.

Data input	1% cpu time.
Geometry, tables	2%
Direct methods (MULTAN etc)	4%
Miscellaneous	8%
Structure factor/Fourier	9%
Structure factor/least-squares	76%

Reading List. Mandatory texts.

Bevington, P.R. (1969) Data Reduction and Error Analysis for the Physical Sciences, McGraw-Hill
Hamilton, W.C. (1964) Statistics in Physical Science, Ronald Press
Lawson, C.L., Hanson, R.J. (1974) Solving Least Squares Problems, Prentice-Hall
Prince, E. (1982) Mathematical Tecniques in Crystallography and Matrials Science, Springer-Verlag.
Rollett, J.S. (1965) Computing Methods in Crystallography, Pergamon

It is impossible to list all important texts.
The following may serve to start a search

Diamond, R. (1981) Structural Aspects of Biomolecules, (ed Srinivasan), Macmillan.
Dunitz, J. (1979) Xray Analysis and the Structure of Organic Molecules, Cornell
Hendrickson, W.A. (1980) Computing in Crystallography (ed Diamond), Indian Academy of Sciences.
Huml, K. (1980) Computing in Crystallography (ed Diamond), Indian Academy of Sciences.
Isaacs, N.(1982) Computational Crystallography (ed Sayre), Clarendon Press
Nicholson, W.L. (1982) Crystallographic Statistics (ed Ramaseshan), Indian Academy of Sciences
Rollett, J.S. (1982) Computational Crystallography (ed Sayre), Clarendon Press

CRYSTALS

David Watkin
Chemical Crystallography Laboratory,
9 Parks Road, Oxford, ENGLAND.

Introduction

Issue 7 is the current version of the program CRYSTALS. The program has been designed to offer powerful and flexible tools for the refinement and evaluation of crystal structures. It contains no direct methods (but is interfaced to MULTAN) nor any molecular graphics (but is interfaced to SNOOPI). The program provides a command-driven user environment with a unified English language interface for interactive, online and batch processing. A 'data-subroutine' facility and a 'pro-forma' data base mean that parameter default values and a data library can easily be provided.

Organisation

The CRYSTALS program is divided into a number of logically distinct levels, each of which is coded with different emphasis. The mathematical sections, which have been coded for numerical accuracy and efficiency, are short and terse though well COMMENTED. The user input sections have been coded to provide an interface that is robust to user error and which can provide substantial support for the novice while retaining power and flexibility for complex problems. The logic involved in these sections is complex. Since this code is only executed for a small fraction of the total running time, clarity is more important than efficiency.

The system has been organised so that the same control structure can be used to run the program in batch, online and in interactive modes, or in a mixture of online and interactive modes during the same terminal session.

The program is command driven, i.e. it doesn't ask questions, but sits waiting for instructions. We adopted this strategy because it is impractical to interrogate the user when there are 78 INSTRUCTIONs, with 327 DIRECTIVEs and 806 keyed parameters. 663 of the DIRECTIVES and parameters have default values. The command driven user environment is like a simplified operating system, in which commands and their data are pre-processed for errors before being executed. The control channel (an input file or the user at the terminal) can pass control to other files, to a depth of 5, control returning to the calling file or master level on exhaustion of the called file or a suitable return statement. This 'data subroutine' structure enables the user or system manager to provide a 'data library', thus avoiding the need to embed data (such as scattering factors or symmetry operators) in the program itself.

During each run of CRYSTALS, all the data needed for the computation is taken from a direct access data base. This file is used for communication between different stages in the computation. It may be created for each run and deleted at the end, or it can be preserved for communication between different runs. Numeric data is presented to CRYSTALS in the form of LISTS, each LIST containing related data. Thus LIST 1 contains the cell dimensions, LIST 2 the symmetry information, LIST 5 the model parameters. The data may either be typed in, or come from a card-image file. It is processed LIST by LIST, checked for syntactic integrity, compared with the pro-forma data base to fill in default values, and then (if accepted) stored in the d.a. file. Users may reinput LISTS at any time, the new LIST either overwriting the previous one, or both being saved. Numeric output from computations is also stored in the d.a. file as LISTS, e.g. the new parameters from a cycle of least squares are stored as a new LIST 5. Extensive facilities exist for checking and manipulating the d.a. file. For example, the user can step back to an old parameter list, or can discard unwanted lists. Some LISTS can be transcribed from the d.a. file to card-image files.

The plain-language output from the program is directed to several channels. The most important of these are a very detailed record of all actions and results in a filestore file for subsequent examination, and a monitor file which gives a sub-set of this information. During online or interactive working, it is this monitoring information which is displayed on the users terminal, providing the facts and evidence needed to proceed safely to the next stage. A 'log' file saves the users input, and can be used as a record of data, or re-used to control futher computations, (even in the current task). The user can examine filestore files (such as data or control files) before executing them and without exiting from CRYSTALS.

The CRYSTALS system provides mechanisms to enable the user to produce simple line drawings of his structure (or part of it) on his terminal, to examine his atomic parameters or results of peak-searches at his terminal, and to modify these at will. A powerful and elegant method of atom or parameter specification enables the user to manipulate his structure via in integral editor, and to accurately control the progress of his refinement.

An extensive set of controls and checks assists the user in monitoring his refinement, and warns him of unusual features. All input to the program is logged, and all data retrieved from the binary data file is dated to avoid ambiguity. In addition, a file accessible only to the system manager records all work done by every user, enabling him to identify users having difficulties or failing to work efficiently.

There is a 230 page User Manual, a 38 page User Guide, various aide-memoires, and various online help facilities.

Features

Atoms are spcified by TYPE and a SERIAL number, e.g. Pb(3). Parameters are specified by a name, e.g. X, U[ISO], or a family name, e.g. X's, U[ij]. Atom parameters are specified by combining these terms, e.g. I(1,X,Z) or Fe(1,X's,U's) - specifying the 3 positional and 6 thermal parameters for Fe(1). Atom parameter values are taken from LIST 5, and may be modified by symmetry operators. C(1) is carbon 1 at its LIST 5 position, S(1,-2,2,1,1,-1) is a sulphur atom at a symmetry equivalent position. The terms -2 to -1 specify the operations, and may be omitted if not required. The special identifiers FIRST and LAST refer to the first and last atoms in LIST 5, and are useful if the LIST has been augmented by a peak search, or re-organised by the molecule assembler. Groups of atoms or parameters may be referenced by the UNTIL sequence, which addresses all those atoms in LIST 5 between and including those specified.

```
C(1) UNTIL H(17) or Zn(2,X'S) UNTIL O(13)
```

Any structural parameter may be modified with the integral editor, and atoms may be selected or rejected by various criteria, e.g.

```
CHANGE SCALE 0.0543
ADD FIRST(Y) UNTIL LAST .25     - shifts the structure to a new origin
```

UEQUIV C(21) UNTIL C(26) - replaces Uaniso by Uequiv for this phenyl
SELECT U[ISO] LT .08

The atom and parameter specifications illustrated above are also used in defining the refinement matrix structure, and the restraints. In a given cycle, the user may define up to 256 matrix blocks, and group which ever parameters he chooses in each block. Parameters may be given either implicitly e.g. X meaning the x coordinates for all atoms, or explicitly, e.g. C(1,X) UNTIL C(6).

BLOCK SCALE DU[ISO] POLARITY EXTPAR — defines a 4x4 block of highly correlated overall parameters
BLOCK X'S — defines a block containing all positions
BLOCK C(1,U'S) UNTIL O(46) — this card with the next puts
PLUS H(11,U[ISO]) UNTIL LAST — the specified temperature factors into a single block

Several physical parameters can be associated with a single l.s. parameter (the derivatives for each parameter summed together), parameter shifts may be linked, and the shifts for equivalenced or linked parameters may be scaled.

EQUIVALENCE C(1,X) C(1,Y) — a single LS parameter coresponding to two physical parameters
EQUIVALENCE Na(1,OCC) K(1,OCC) — this card, with the next, applies
WEIGHT -1 K(1,OCC) — equal but opposite shifts to the site occupation factors
LINK C(1,X'S) H(11,X'S) UNTIL H(13) — a riding Methyl group
LINK C(11,U'S) UNTIL C(16) — a group anisotropic thermal parameter

For more complex cases, the user may build each block element by element using individual parameter specifications. Matrix accumulation is double precision, with Choleski inversion. Near singularities cause the appropriate parameter shifts to be set to zero. The normal matrix may be stored for re-use with an updated vector. A and B parts of the structure factors can be saved for those atoms not being refined. The computed shifts can be damped or augmented, and a wide range of weighting schemes, including observed weights, Hughes, Cruickshank or Dunitz weights, and automatic re-weighting, is provided. Refinement can be against F or F**2, and data for twinned crystals with up to 9 components can be handled. The refinable parameters are :

x,y,z, U[iso], U[11] etc., occupation factor, overall scale, layer and batch scales, Larsen extinction factor, dummy U[iso], f'', absorption function.

A good starting model can be obtained by automatic Fourier refinement, regularisation (with either an internal or an external model), and geometric positioning.

The restraints available include planarity, mean or sum of parameters, distances and angles either to defined values or to means or differences, terms of U[ij], and vibrations along vectors. The user may also provide a FORTRAN like expression in the crystallographic parameters to define his own restraint, which is differentiated numerically.

SUM Y	restrains the sum of the y shifts, e.g. to fix the origin along a polar axis
PLANAR C(1) UNTIL C(6)	holds the phenyl planar
VIB 0,.005 = C(1) TO C(2)	vibration restraint
DIST 1.390,.002 = C(1) TO C(1,-1)	restrains a distance across a centre of symmetry

Facilities are available for distances and angles, torsion angles, best lines and planes, principal axes of Uaniso, TLS analysis, fragment comparing, and for producing geometry, parameter and reflection tables for publication. Simple line plots of the structure can be produced on a dumb terminal or printer.

CRYSTALS is an evolving system, and occasionally metamorphoses

'EDITION 1'	(Data base definition) (Cruickshank, Freeman, Rollett, Truter, Sime, Smith and Wells, 1964)
'NOVTAPE'	(AUTOCODE) (Hodder, Rollett, Prout and Stonebridge, Oxford, 1964),
'FAXWF'	(ALGOL) (Ford and Rollett, Oxford, 1967),
'CRYSTALS'	(FORTRAN) (Carruthers and Spagna, Rome, 1970)
'CRYSTALS'	(FORTRAN) (Carruthers, Prout, Rollet and Spagna, Oxford, 1975)
'CRYSTALS'	Issue 2 (FORTRAN) (Carruthers, Prout, Rollet and Watkin, Oxford, 1979)
'CRYSTALS'	Issue 7 (FORTRAN) (Betteridge, Prout and Watkin, Oxford, 1984)

ABSOLUTE STRUCTURE AND HOW NOT TO DETERMINE IT

by Peter G. Jones

Institut für Anorganische Chemie der Universität, Tammannstrasse 4,
D-3400 Göttingen

I. Absolute configuration and absolute structure

Since the pioneering work of Bijvoet it has been well known that anomalous dispersion effects [associated with the breakdown of Friedel's law $I(hkl) = I(\bar{h}\bar{k}\bar{l})$] in non-centrosymmetric structures can be exploited to distinguish such a structure from its inverse and thus to determine ... well, what? Absolute configuration, absolute conformation, polar axis direction, chirality, enantiomorph, enantiomorphous space group - the list is long and confusing.

The initial application of anomalous scattering was to the determination of the absolute configuration of sodium rubidium tartrate. This chiral (optically active) material contains two asymmetric centres, the configurations of which can be described by the Cahn-Ingold-Prelog *R*,*S* notation (in fact both asymmetric carbon atoms are *R*); and the corresponding mirror-image molecules (with *S*,*S* configuration - the opposite enantiomer) are not present in the structure. This example demonstrates the minimum requirements to justify a claim to an absolute configuration: the structure must contain at least one asymmetric centre and only one enantiomer.

Conversely, an absolute configuration cannot be determined (because there is no such thing) in the following cases: (i) non-centrosymmetric but achiral space groups where the symmetry operations include some form of mirror-image generation, e.g. $Pna2_1$ with glide planes, (ii) molecules lacking asymmetric centres (achiral molecules), even if they crystallize in chiral space groups. The question thus arises, what *is* being determined as the manifestation of non-centrosymmetry in these cases?

Absolute conformation (the sign of all torsion angles) is determined for achiral molecules in chiral space groups; inverting the structure changes the sign of all torsion angles.

For most achiral non-centrosymmetric space groups, the polar axis direction is determined. Unfortunately, the term "polar" has been used to mean many different things, but current usage seems to be confined to space groups with a floating origin in one or more direction(s) - such a direction is then known as polar axis, and is inverted in the inverse structure.

(A polar axis can easily be identified by inspection of symmetry operators; thus if no symmetry operation causes the sign of, say, *y* to be negative then the *y* axis is polar, e.g. in $P2_1$ with operators *x*, *y*, *z* and -*x*, 0.5+*y*, -*z*. This also shows that both chiral and achiral non-centrosymmetric space groups may possess polar axes).

There are however achiral non-centrosymmetric space groups lacking polar axes, eg $I\bar{4}$. In this case it may be considered that the absolute assignment of x and y axes is determined, since changing x to -y and y to x also changes the set of symmetry operators to their inverse. For some enantiomorphous pairs of space groups (eg $P3_121$ and $P3_221$) an absolute configuration (or conformation) is determined simultaneously with the space group itself.

In view of all these complications, it is unfortunate that no general term exists to describe the process of distinguishing a non-centrosymmetric structure from its inverse. I should like to suggest "determination of absolute structure".

II. The determination of absolute structure: data collection strategy

In addition to the usual goals of maximum precision and minimum systematic errors, the intensity differences between *hkl* and $\overline{h}\overline{k}\overline{l}$ should be maximised. A long wavelength radiation may be more appropriate (although a common exception is furnished by Br, which shows a much higher f''_i for Mo *K*α than for Cu *K*α) and $2\theta_{max}$ should be as high as is consistent with the scattering power of the crystal since (a) the heavy-atom scattering, which includes the major anomalous scattering, dominates at high 2θ, and (b) *f*" is insensitive to θ, thus becoming more important relative to *f* at high θ. It should go without saying that Friedel opposites should be collected and, for extremely accurate work, several sets of equivalents also. Since the presence of good anomalous scatterers is usually associated with moderate to severe absorption, particularly if Cu *K*α radiation is used, and since absorption effects are generally much greater than the small Friedel differences, an absorption correction is often essential. Finally, measuring *hkl* at 2θ, ω, χ, ϕ followed immediately by $\overline{h}\overline{k}\overline{l}$ at -2θ, ω-2θ, χ, ϕ will minimise not only errors associated with differing path lengths (given a crystal with centrosymmetric shape) but also errors caused by crystal decay or medium term fluctuations in primary beam intensity.

The datasets are solved and refined in the usual way and the subjected to one or more of the discriminatory tests such as Hamilton's *R* test (1965), Rogers' η refinement (1981) and Flack's *x* refinement (1983).

The latter two methods have two main advantages over Hamilton's method; they provide a measure of confidence (the least-squares e.s.d. of η or $\underline{x}$) in the assignment of absolute structure and they are valid in all non-centrosymmetric space groups, even those (such as Fdd2) in which inversion in the origin does not generate the alternative absolute structure. Unfortunately the η and $\underline{x}$ methods have not yet gained universal acceptance. (See also the article "Absolute Configuration Determination" by H.D. Flack).

In borderline cases (Rabinovich & Hope, 1980) it may be necessary to calculate which reflections are most sensitive to anomalous scattering effects and to remeasure them extremely accurately.

III. Published Absolute Configurations, 1982

In 1979 Rogers & Allen noted several unsatisfactory features regarding publication of absolute configurations; it seemed worthwhile to inspect more recent reports to see if the situation had improved. A search of the Cambridge Crystallographic Data Centre files for 1982 provided 102 "absolute configuration" flags. Detailed inspection of these revealed few grounds for optimism.

Data collection. (i) Friedel opposites. There were 72 structures for which some form of refinement was used to determine absolute configuration. Only 14 datasets contained a full set of Friedel opposites; in the absence of Friedel opposites, a low $2\theta_{max}$ led in many cases to a very low data/parameters ratio, eg 588 reflections for 32 atoms. (ii) Absorption corrections. Since good anomalous scatterers usually absorb well, an absorption correction should normally be performed. However, of 38 materials with $\mu > 3$ mm^{-1}, only 12 were definitely corrected for absorption. Can the small Friedel differences be reliably measured in such cases (eg μ = 15.3 mm^{-1})?

Achiral structures. Two structures in the achiral space group $Pna2_1$ were assigned an "absolute configuration"; "absolute structure" would have been an appropriate term. Further, an absolute configuration was also claimed for one molecule with "virtual m symmetry".

Over-optimism. Ten structures showed a high R value (0.07 to 0.11); can Friedel differences at this level of accuracy be reliable? A further eleven cases showed a difference in R values (between alternative structures) < 0.1 % which, in view of doubts cast on the validity of Hamilton's test in borderline cases by Rogers (1981) & Marsh (1981), must be considered of limited significance despite the calculated probabilities (which neglect the possibility of systematic errors).

Presentation of results. The modern tendency to abbreviation of crystallographic information led in one case to the relegation of all such material to a deposited Supplement. Nine further publications gave either no details of the method used to determine absolute configuration or merely a brief mention of "use of anomalous dispersion".

IV. Recommendations for Determination & Publication of Absolute Structure

(i) In all cases where a structure contains anomalous scatterers and is known to be non-centrosymmetric, Friedel opposites should be collected. (ii) Absorption corrections should be applied if absorption effects are likely to be significant (perhaps for $\mu t > 0.3$, as a rule of thumb, although crystal shape is also an important factor). (iii) Results of statistical tests should be interpreted with caution. (iv) A clear statement that an absolute structure has been determined should be alloted a prominent place in any publication. (v) Details of data collection (in particular $2\theta_{max}$, μ and information on Friedel opposites and absorption corrections), and of the method used to assign the absolute structure, should be given. (vi) Absolute configuration should not be claimed for achiral molecules or space groups. (vii) Referees and editors must adopt a more critical attitude.

V. Neglect of Anomalous Scattering

If a chemically important compound is inconsiderate enough to crystallize in a non-centrosymmetric space group, and its absolute structure is not of primary interest, little or no effort may be made to assign an absolute structure. The consequences, in terms of possible large systematic errors in heavy atom positions, have been made clear by Cruickshank & MacDonald (1967). A survey of recent non-centrosymmetric structures not bearing the Cambridge flag "absolute configuration" will be carried out in the near future in order to check for such errors.

References

CRUICKSHANK, D.W.J. & MCDONALD, W.S. (1967). Acta Cryst. 23, 9-11.

FLACK, H.D. (1983). Acta Cryst. A39, 876-881.

HAMILTON, W.C. (1965). Acta Cryst. 18, 502-510.

MARSH, R.E. (1981). Acta Cryst B37, 1985-1988.

RABINOVICH, D. & HOPE, H. (1980). Acta Cryst. A36, 670-678.

ROGERS, D. & ALLEN, F.H. (1979). Acta Cryst. B35, 2823-2825.

ROGERS, D. (1981). Acta Cryst. A37, 734-741.

A greatly expanded version of Sections I-IV of this article has been accepted for publication in Acta Crystallographica A under the title "The Determination of Absolute Structure (Parts I & II)".

COMPUTATIONAL MODELS OF CRYSTALS*

William R. Busing
Chemistry Division, Oak Ridge National Laboratory,
Oak Ridge, Tennessee 37831, USA

Computational models of crystal structures have been used extensively to interpret and predict the atomic or molecular arrangements and some properties of crystalline substances. Work on molecular crystals has been summarized by Timofeeva, Chernikova, & Zorkii (1980). Included are publications by Warshel & Lifson (1970), Kitaigorodsky (1973), Momany, Carruthers & Scheraga (1974), Gavezzotti & Simonetta (1975), Williams & Starr (1977), Taddei, Righini & Manzelli (1977), and Busing (1982, 1983). Application of these methods to inorganic crystals includes work by Born & Mayer (1932), Tosi & Fumi (1964), Slaughter (1966), Busing (1970, 1972a), Ohashi & Burnham (1972), Giese & Datta (1973), Catlow & Norgett (1973), Yuen, Murfitt, & Collin (1974), Brown & Fenn (1979), Miyamoto & Takeda (1980), Matsui & Watanabe (1981), Matsui & Matsumoto (1982), Catlow & Parker (1982), and Matsui & Busing (1984a,b).

Computer programs for this kind of calculation include PCK5 and PCK6 (Williams, 1969, 1972, 1979), QCFF/PI and MCA (Warshel & Levitt, 1982; Huler, Sharon & Warshel, 1977; Warshel, 1977), WMIN (Busing, 1972b, 1981), and PLUTO and METAPOCS (Catlow & Mackrodt, 1982).

A computational model consists of a description of the crystal structure, a procedure for calculating the potential energy of a given structure, and a way of minimizing the calculated energy by adjusting the structural variables. The structure description includes the lattice parameters, the coordinates of the atoms in one asymmetric unit, and the symmetry operations which can be used to generate the entire structure.

The energy calculation often uses the atom-atom approximation which includes a term $V(r_{ij})$ for each nonbonded pair separated by distance r_{ij}. This potential is often written

$$V(r_{ij}) = q_i q_j r_{ij}^{-1} - A r_{ij}^{-6} + B \exp(-C r_{ij}), \qquad (1)$$

where the terms represent Coulomb, van der Waals, and repulsion energy,

*Research sponsored by the Division of Materials Sciences, U. S. Department of Energy under contract DE-AC05-84OR21400 with the Martin Marietta Energy Systems, Inc.

respectively. The atomic charges q_i and the coefficients A, B, and C, which depend on the kinds of atoms involved, are known as the nonbonded energy parameters.

When molecules are treated as flexible groups it is necessary to include terms describing the intramolecular energy in some form such as

$$V(r) = (k_r/2)(r - r_0)^2 , \qquad (2)$$

$$V(\alpha) = (k_\alpha/2)(\alpha - \alpha_0)^2 , \qquad (3)$$

$$V(d) = (k_d/2)\, d^2, \qquad (4)$$

$$V(\phi) = (E_\phi/4)(1 - \cos^2\phi). \qquad (5)$$

Here r is a bond distance, α a bond angle, d is an out-of-plane distance, and ϕ is a torsion angle. The energy parameters include the force constants k_r, k_α, and k_d, the unstrained values r_0 and α_0, and the 2-fold barrier to torsion E_ϕ.

X-ray crystallography is the most precise and unambiguous technique for observing the geometry of molecules, but molecules in crystals are always distorted to some extent by intermolecular packing forces. In general these nonbonded forces are well understood, in that the energy parameters of (1) are known (Williams & Starr, 1977; Mirsky, 1978). Balancing these packing forces is the intramolecular potential (2) to (5) with parameters which can be determined by adjusting them until the calculated structure reproduces that observed (Williams, 1972; Busing, 1983). The potential can then be used to calculate the geometry of an isolated molecule. The computational model thus provides a means of going from the molecular geometry observed in the crystal to the geometry inherent in the molecule itself.

As an example of the usefulness of this procedure consider Fig. 1 which illustrates the molecular geometries of 1,8-diphenylnaphthalene (DPNAP) and 1,4,5,8-tetraphenylnaphthalene (TPNAP) as observed in the crystal (Ogilvie, 1971; Evrard, Piret & Van Meerssche, 1972). The torsion angles listed on the figure indicate that, although the naphthalene nucleus in DPNAP is nearly planar, in TPNAP it is highly distorted. Clough, Kung, Marsh & Roberts (1976) proposed a molecular explanation for this difference, namely that in DPNAP the distorting forces of the phenyl groups can be accommodated by in-plane changes of the naphthalene

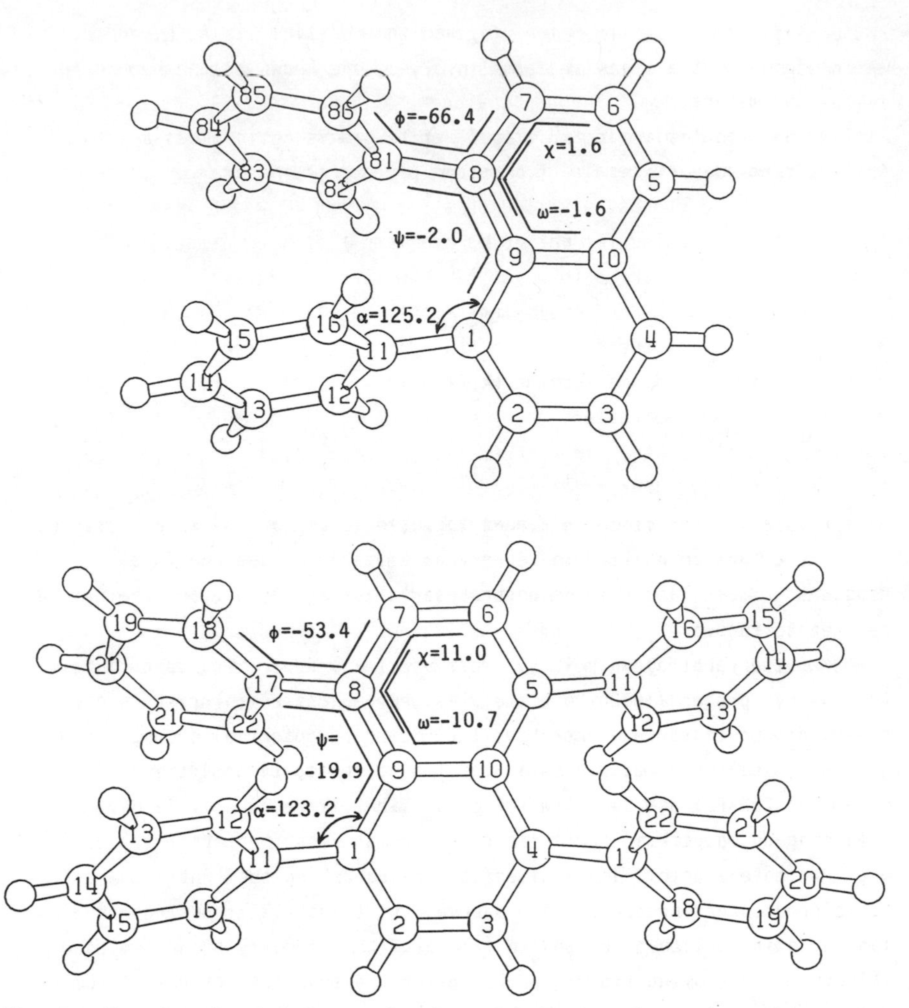

Fig. 1. Two phenylnaphthalene molecules with bond angles and torsion angles (degrees) observed in the crystal structures.

distances and angles. In TPNAP, with phenyl groups on both sides of the naphthalene, such in-plane changes cannot relieve the strain, and out-of-plane distortions are required. Because the packing arrangements are quite different in these two crystals, it is possible that crystal packing forces play an important role as the origin of this difference in geometry. To test this hypothesis we modeled the two compounds, adjusting the intramolecular energy parameters to reproduce the observed crystal structures. Then the molecular geometries of the isolated molecules were calculated and compared.

In the model for DPNAP the 10 carbon and 6 hydrogen atoms of the naphthalene nucleus were treated as independent, each with 3 degrees of translational freedom. The naphthalene-phenyl links, 1-11 and 8-81, were treated as rigid, pivoting on atoms 1 and 8, each with 2 degrees of rotational freedom. The remaining 10 atoms of each phenyl radical were treated as rigid groups, pivoting on atoms 11 and 81, each with 3 rotations. These 58 molecular variables and the 4 lattice parameters of the monoclinic crystal make a total of 62 structural variables.

The model for TPNAP is similar, but this molecule is located on an inversion center of the crystal. The asymmetric unit requires only 5 carbon and 2 hydrogen atoms for the naphthalene nucleus, each with 3 degrees of freedom. Two naphthalene-phenyl links and 2 rigid phenyl groups are required, as for DPNAP. The other half of the molecule is generated by the center of inversion. There are 31 internal variables and 4 parameters of the monoclinic lattice, making a total of 35 structural variables.

Energy was calculated using expression (1) with parameters from Williams & Starr (1977) for the nonbonded interactions. Bond stretching potentials (2) were used to describe 2 kinds of C-C bonds and the C-H bonds of the naphthalene moiety. Parameters k_r and r_0 were taken from Allinger & Sprague (1973), and the r_0's were adjusted to reproduce the observed bond distances. One bond angle bending potential (3) was used with $\alpha_0 = 120°$; parameter k_α was adjusted to reproduce angles 9-1-11 and 9-8-81 in DPNAP and 9-1-11 and 9-8-17 in TPNAP. One out-of-plane potential (4) and one torsion potential (5) tend to keep the naphthalene and the naphthalene-phenyl links planar. Parameters k_d and E_ψ were adjusted to reproduce torsion angles ψ, χ, and ω. Finally, a second torsion potential (5) was applied to establish the phenyl-naphthalene torsion angles ϕ. The parameter E_ϕ required to reproduce the observed angle ϕ was slightly different for the two compounds. All other energy parameters were the same for both materials.

Table 1 shows a comparison of several parameters of the observed and calculated structures for the two substances. The agreement is quite good, and this demonstrates that a model which is almost the same for the two molecules is capable of reproducing the observed structures, in spite of the considerable difference between their geometries.

Also shown in Table 1 are the torsion angles calculated for the isolated molecules by using the same model. The degree of distortion of the naphthalene nucleus is predicted to be about the same for the two compounds. Thus it appears that crystal packing forces enhance the

Table 1. Observed and calculated structural parameters for phenylnaphthalenes

	DPNAP			TPNAP		
	Crystal		Isolated molecule	Crystal		Isolated molecule
	Obs	Calc	Calc	Obs	Calc	Calc
a/Å	8.58	8.41		6.46	6.46	
b	20.03	19.84		24.33	24.19	
c	9.72	9.66		8.02	7.90	
β/°	116.7	115.7		114.3	114.1	
φ	-66.4	-66.5	-52.8	-53.4	-52.3	-62.5
ψ	-2.0	-2.3	-16.4	-19.9	-20.5	-14.4
χ	1.6	2.2	4.3	11.0	10.9	6.8
ω	-1.6	-1.7	-8.5	-10.7	-10.7	-6.6
α	125.2	125.1	123.9	123.2	123.4	124.5

distortion of the naphthalene moiety in TPNAP, but in DPNAP they actually make it more planar. Further calculations make it clear that the processes proposed by Clough, Kung, Marsh & Roberts (1976) play only a small part in determining these geometries, and the effects of crystal packing forces predominate.

Recently Busing & Matsui (1984) described a general method for applying external forces such as hydrostatic pressure, normal or shear stresses, and electric fields to the computational model of a crystal. The method involves adding to the calculated crystal energy an extra term, W_{ext}, which is the energy of the external force-producing device. When the total energy of the system is minimized the crystal structure adjusts itself to take account of the added force. For example, to apply a normal tension σ_1 along the a-axis of an orthorhombic crystal we add the term

$$W_{ext} = -\sigma_1 bc(a - a_0)/Z \qquad (6)$$

to the energy per formula unit. When the total energy is minimized, the equilibrium lattice parameter, a, will become larger than the zero-stress value a_0. In response to the crystal forces, b and c will become smaller. The computed strains are $\varepsilon_1 = \Delta a/a$, $\varepsilon_2 = \Delta b/b$, and $\varepsilon_3 = \Delta c/c$; and the elastic compliance constants, s_{ij}, can be calculated, since $\varepsilon_1 = s_{11}\sigma_1$, $\varepsilon_2 = s_{12}\sigma_1$, and $\varepsilon_3 = s_{13}\sigma_1$. Other compliance constants can be obtained by applying stresses σ_2 and σ_3 along b and c, respectively.

Busing & Matsui (1984) showed how this procedure can be used to apply either normal or shear stresses to crystals of any symmetry. They described how the crystal symmetry must be relaxed in some cases. Ways of extrapolating the elastic constants to their values at zero stress were also discussed.

In the modeling of minerals, Matsui & Busing (1984a) emphasized the importance of reproducing both the observed crystal structures and the experimental elastic constants. Two early models for forsterite, the olivine form of Mg_2SiO_4, (Miyamoto & Takeda, 1980; Matsui & Matsumoto, 1982) produced good structures, but calculation of the elastic constants showed that these models were much too stiff.

Matsui & Busing (1984b) made similar calculations on the pyroxene mineral diopside, $CaMgSi_2O_6$. Fig. 2 illustrates this crystal structure (Warren & Bragg, 1928; Levien & Prewitt, 1981; Rossi, Ghose & Busing, 1982). The crystal is monoclinic with space group C2/c and four $CaMgSi_2O_6$ units per cell. The Ca and Mg ions each occupy sites on 2-fold axes parallel to b (perpendicular to the page in the figure). One kind of Si atom and three kinds of O atoms are located in general positions. The O atoms are arranged about the Si atoms to form tetrahedral SiO_4 groups, and these are linked to each other to form one-dimensional chains parallel to the c axis. Each tetrahedral group is related to its neighbors in the chain by the c-glide plane, which is parallel to the page in the figure.

These silicate chains are stacked together to form layers parallel to the b,c plane. The silicate layers are held together by intermediate layers of cations. Each Mg ion is octahedrally coordinated to 6 oxygen atoms of the silicate chain, and each Ca is coordinated less regularly to 8 oxygen atoms.

Three different models were tried for the silicate chain. The most successful of these permitted all bond angles and the Si-O3 bond distances to vary. Structural variables include 10 degrees of freedom for the chain, 2 cation coordinates, and 4 monoclinic lattice parameters for a total of 16.

The nonbonded interaction potential (1) used only Coulomb and repulsion terms. Bond stretching and bond-angle bending contributions, (2) and (3), were included for the silicate chain. Repulsion parameters were fixed at values derived previously for the olivine structures.

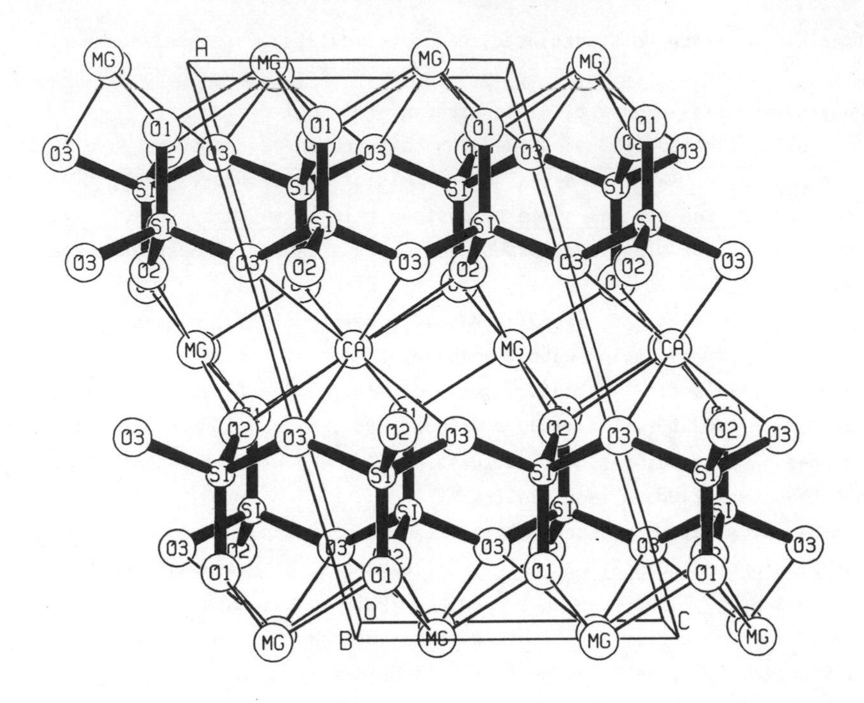

Fig. 2. The crystal structure of diopside. Pairs of one-dimensional silicate chains overlap each other in this view down the b axis.

Charges on the Ca and Mg ions were fixed at 2 proton units. Partial charges of the O and Si atoms and parameters of the bond stretching and bond-angle bending potentials were adjusted systematically to develop a model which reproduced both the crystal structure and the observed elastic constants reasonably well. Table 2 lists the observed and calculated lattice parameters, mean interatomic distances, and the Si-O3-Si bond angle. Although the calculated b is 5.7% too large, the other lattice parameters are off by less than 1.5%, and the β angle is correct to 0.3°. Mean interatomic distances are good to 0.05 Å, but the Si-O3-Si angle is 4.5° too large. Table 3 lists the observed (Levien, Weidner & Prewitt, 1979) and calculated elastic constants. The greatest discrepancy is in the cross term c_{23} which calculates 26% too large.

Table 2. Observed and calculated parameters of the crystal structure of diopside. Distances in Å and angles in degrees.

Parameter	Obs	Calc
a	9.75	9.60
b	8.92	9.43
c	5.25	5.28
β	105.9	106.2
Si-O3-Si	135.8	140.3
<Si-O>	1.64	1.63
Si-Si	3.11	3.15
<Mg-O>	2.08	2.12
<Ca-O>	2.50	2.55

Table 3. Observed and calculated elastic constants and bulk modulus K for diopside, Mbar.

ij	Obs c_{ij}	Calc c_{ij}
11	2.23	2.10
22	1.71	1.66
33	2.35	2.42
44	0.74	0.80
55	0.67	0.70
66	0.66	0.58
12	0.77	0.64
13	0.81	0.87
23	0.57	0.72
15	0.17	0.26
25	0.07	0.11
35	0.43	0.49
46	0.073	0.092
K	1.08	1.05

Having established that this model for diopside is reasonable, Matsui & Busing (1984b) proceeded to calculate the structure and elastic constants under 50 kbar hydrostatic pressure. Changes in the calculated structural parameters agreed approximately with the changes observed experimentally by Levien & Prewitt (1981), who studied diopside at 5 different pressures. For the calculated pressure derivatives of the elastic constants, only that of the bulk modulus K is available from experiment. The calculated value of $\partial K/\partial p$ is 6.2 compared with the value of 4.8 reported by Levien, Weidner & Prewitt (1979), so this shows that the calculated derivatives are at least of the right order of magnitude.

Busing & Matsui (1984) also discussed the application of electric fields and the calculation of dielectric and piezoelectric coefficients. By applying more than one kind of force simultaneously, higher order elastic constants and derivatives of the piezoelectric constants can be obtained. It should be possible to simulate second-order phase transitions which occur under the influence of pressure. The same should be true of ferroelastic transitions, which may be induced by shearing stresses, or of ferroelectric transitions, which occur under the influence of an electric field. It should also be possible to predict the effect of forces or fields on nonlinear optical properties which depend on the geometry and orientation of molecules in crystals.

Thus we have seen two general ways in which computational models of crystals can be used to extend the information available from experiments. The calculation of intramolecular potentials and the geometry of isolated molecules from observed crystal structures provides knowledge which is presently unavailable from other sources. The extrapolation of crystal structures and properties to high pressures should provide a guide for experiments which are only just beginning to be undertaken. It is clear that new uses for these models will continue to be found.

References

Allinger, N. L. & Sprague, J. T. (1973). J. Am. Chem. Soc. 95, 3893-3907.

Born, M. & Mayer, J. E. (1932). Z. Phys. 75, 1-18.

Brown, G. E. & Fenn, P. M. (1979). Phys. Chem. Minerals 4, 83-100.

Busing, W. R. (1970). Trans. Amer. Cryst. Assoc. 6, 57-72.

Busing, W. R. (1972a). J. Chem. Phys. 57, 3008-3010.

Busing, W. R. (1972b). Acta Cryst. A28, S252.

Busing, W. R. (1981). WMIN. Report ORNL-5747. Oak Ridge National Laboratory, Oak Ridge, TN.

Busing, W. R. (1982). J. Amer. Chem. Soc. 104, 4829-4836.

Busing, W. R. (1983). Acta Cryst. A39, 340-347.

Busing, W. R. & Matsui, M. (1984). Acta Cryst. A40. In press.

Catlow, C.R.A. & Mackrodt, W. C. (1982). In Computer Simulation of Solids, edited by C.R.A. Catlow and W. C. Mackrodt, pp 3-20. New York: Springer-Verlag.

Catlow, C.R.A. & Norgett, M. J. (1973). J. Phys. C. 6, 1325-1339.

Catlow, C.R.A. & Parker, S. C. (1982). In Computer Simulation of Solids, edited by C.R.A. Catlow and W. C. Mackrodt, pp 222-240. New York: Springer-Verlag.

Clough, R. L., Kung, W. J., Marsh, R. E. & Roberts, J. D. (1976). J. Org. Chem. 41, 3603-3609.

Evrard, G., Piret, P. & Van Meerssche, M. (1972). Acta Cryst. B28, 497-506.

Gavezzoti, A. & Simonetta, M. (1975). Acta Cryst. A31, 645-654.

Giese, R. F., Jr. & Datta, P. (1973). Am. Mineral. 58, 471-479.

Huler, E., Sharon, R. & Warshel, A. (1977). QCPE 11, 325.

Kitaigorodsky, A. I. (1973). Molecular Crystals and Molecules. New York: Academic Press.

Levien, L. & Prewitt, C. T. (1981). Am. Mineral. 66, 315-323.

Levien, L., Weidner, D. J. & Prewitt, C. T. (1979). Phys. Chem. Minerals 4, 105-133.

Matsui, M. & Busing, W. R. (1984a). J. Phys. Chem. Minerals. In press.

Matsui, M. & Busing, W. R. (1984b). Am. Mineral. In press.

Matsui, M. & Matsumoto, T. (1982). Acta Cryst. A38, 513-515.

Matsui, M. & Watanabé, T. (1981). Acta Cryst. A37, 728-734.

Mirsky, K. (1978). Computing in Crystallography, edited by H. Schenk, R. Olthof-Hazekamp, H. van Koningsveld & G. C. Bassi, pp 169-182. Delft University Press: Delft.

Miyamoto, M. & Takeda, H. (1980). Geochem. J. 14, 243-248.

Momany, F. A., Carruthers, L. M. & Scheraga, H. A. (1974). J. Phys. Chem. 78, 1621-1630.

Ogilvie, R. A. (1971). Ph.D. Thesis, Mass. Inst. of Tech.

Ohashi, Y. & Burnham, C. W. (1972). J. Geophys. Res. 77, 5761-5766.

Rossi, G., Ghose, S. & Busing, W. R. (1982). Geol. Soc. Am. Abstr. 14, 603.

Slaughter, M. (1966). Geochim. Cosmochim. Acta 30, 315-322.

Taddei, G., Righini, R. & Manzelli, P. (1977). Acta Cryst. A33, 626-628.

Timofeeva, T. V., Chernikova, N. Yu. & Zorkii, P. M. (1980). Uspekhi Khimii 49, 966-997; Russian Chemical Reviews 49, 509-525.

Tosi, M. P. & Fumi, F. G. (1964). J. Phys. Chem. Solids 25, 45-52.

Warren, B. & Bragg, W. L. (1928). Z. Krist. 69, 168-193.

Warshel, A. (1977). Comp. Chem. 1, 195-202.

Warshel, A. & Levitt, M. (1982). QCPE 11, 247.

Warshel, A. & Lifson, S. (1970). J. Chem. Phys. 53, 582-594.

Williams, D. E. (1969). Acta Cryst. A25, 464-470.

Williams, D. E. (1972). Acta Cryst. A28, 629-635.

Williams, D. E. (1979). QCPE 11, 373.

Williams, D. E. & Starr, T. L. (1977). Comp. Chem. 1, 173-177.

Yuen, P. S., Murfitt, R. M. & Collin, R. L. (1974). J. Chem. Phys. 61, 2383-2393.

Computer methods in protein crystallography

ANOMALOUS DISPERSION IN PHASE DETERMINATION FOR MACROMOLECULES

Wayne A. Hendrickson

Department of Biochemistry and Molecular Biophysics
Columbia University, New York, NY 10032, U.S.A.

and

Laboratory for the Structure of Matter
Naval Research Laboratory, Washington, D. C. 20375, U.S.A.

The scattering of x-rays by matter arises from the acceleration of electrons in vibrations excited by the incident x-ray wave. The theoretical scattering from a free electron does not depend upon the wavelength of the radiation. However, resonance between induced oscillations and the natural frequencies of vibration of bound electrons in atoms will be expected to modulate the x-ray scattering and make it wavelength dependent. The resulting dispersive property is called anomalous scattering, or anomalous dispersion, although it is actually the norm for all but the lightest of atoms when one considers the wavelengths of interest in x-ray diffraction experiments. The effects increase sharply as the x-ray energy approaches an atomic absorption edge and can be especially large for certain ionic species.

The potential usefulness of anomalous scattering in structure analysis was first recognized by Bijvoet (1949). He suggested that departures from Friedel's law could be used to determine the absolute configuration of handed molecules and that anomalous scattering could also be used to resolve the phase ambiguity in the single isomorphous replacement experiment. Methodology for realizing the phasing potential of anomalous scattering in the isomorphous replacement method was worked out early in protein crystallography and the use of anomalous scattering as an adjunct to isomorphous replacement is now a standard of the field. The proper conjunction of these two sources of phase information gives the structure in the correct hand as a happy by-product. Anomalous scattering has also been used very effectively to locate native metal centers in protein crystals. Such information can be extremely useful in the course of phase determination by molecular replacement from a similar known structure.

These conventional uses of anomalous scattering -- to determine the absolute configuration, to find the positions of native metal centers, and to supplement isomorphous replacement phasing -- have considerable importance in macromolecular crystallography. However, more recent developments are of greater interest here as they are concerned with the use of anomalous scattering for the direct determination of phase angles. That is, direct in the sense that only the data from a single crystalline species are required. Three major analytic procedures have been implemented or considered for direct phasing from anomalous scattering. One of these uses the normal scattering contributions from the anomalous scatters to resolve the ambiguity inherent in phase information from anomalous scattering at a single wavelength. Several structures have now been solved in this way. The second major category of direct anomalous phasing is based on measurements at multiple wavelengths. Such measurements can be made with the tunable radiation from synchrotron sources and can in principle lead to a definitive solution of the phase problem. Experimental work is in progress at several centers. The final class of procedures involves the integration of classical direct methods with anomalous scattering measurements. The formulations show powerful promise, but practical implementation has not been reported.

In this article I will indicate the theoretical basis for the various approaches to phase determination that involve anomalous scattering and then describe some of the computational procedures used in two of these approaches. No attempt will be made to review the literature or to describe particular applications.

Basic Relationships

The distinctiveness of an anomalous scattering center can be expressed in its atomic scattering factor. The normal scattering factor, f°, that would pertain in the absence of resonance effects is modified by a wavelength dependent correction -- the anomalous dispersion -- that involves a phase shift δ with respect to the normal scattering factor. Thus,

$$f = f^\circ + f^{\Delta}e^{i\delta} = f^\circ + \Delta f' + i\Delta f'' \quad . \qquad (1)$$

The most direct impact of the imaginary component, $\Delta f''$, is seen in a breakdown of Friedel's law of symmetric diffraction. Effects of the real component, $\Delta f'$, are evident in intensity differences at different wavelengths. The manner in which these effects are exploited in phase determinations depends on the application.

Isomorphous replacement. Anomalous scattering is appreciable for nearly all atoms heavy enough to be useful isomorphous replacement for macromolecules. Accordingly, anomalous scattering is generally available as a auxiliary source of phase information and it has the powerful advantage of being complementary in nature. The diffraction pattern of a crystal of native molecules is given by structure factors F_P; that of the derivative complex has structure factors F_{PH} to which the array of heavy atoms contributes F_H. In the absence of measurement errors or lack of isomorphism, these are related by $F_{PH} = F_P + F_H$ and it follows that

$$|F_{PH}|^2 - |F_P|^2 = |F_H|^2 + 2|F_P||F_H| \cos (\phi_P-\phi_H). \quad (2)$$

Thus, in principle, the protein phase ϕ_P must in general have one of two values if $|F_{PH}|$, $|F_P|$ and $F_H = |F_H| \exp (i\phi_H)$ are known. This ambiguity can be resolved by other derivatives or, in principle, by the anomalous scattering. One can readily demonstrate that, in the case of only one kind of anomalous scatterer and with perfect data,

$$|F_{PH}(h)|^2 - |F_{PH}(-h)|^2 = 4 \; |F_P||F_H''| \sin (\phi_P-\phi_H) \quad (3)$$

where $|F_H''|$ is the contribution from the imaginary component of scattering and h is the vector of Miller indices. The trigonometric complementarity in the isomorphous and anomalous differences is evident in equations (2) and (3). The combination of information leads to good estimates of $|F_H|$ from which the heavy atom positions can be found as shown by Matthews (1966), for example, and then the phases ϕ_H can be calculated to yield ϕ_P from the preceding relationships (North, 1965).

Resolved anomalous phasing. It often happens that native biological macromolecules contain significant anomalous scatterers or that such centers are introduced during crystallization. We have recently shown in the structure determination for crambin (Hendrickson and Teeter, 1981) that if these anomalous scattering centers also constitute a substantial partial structure, then a direct phase determination is possible. The general relationship that applies is

$$|F(h)|^2 - |F(-h)|^2 = 4 \; |F'| \; |F_A''| \cos (\psi+\omega-\phi). \quad (4)$$

Here $F' = |F'| \exp(i\psi)$ is the structure factor due to the real parts

($f' = f° + \Delta f'$) of the scattering and $F''_A = |F''_A| \exp[i(\psi+\omega)]$ is that from the imaginary components of the set of anomalous scattering centers.

In the event that all anomalous scatterers of of the same atomic species and the anomalous scattering is weak relative to the total scattering, (4) reduces to

$$\Delta F = |F(h)|-|F(-h)| \simeq -2\,|F''_A| \sin(\psi-\phi). \qquad (5)$$

This relationship can be used to deduce and refine the atomic positions of the anomalous scatterers and thereby obtain the phase ψ. The information about ϕ is ambiguous and, due to errors, imprecise. However, the phase information can usefully be expressed as a probability distribution (Hendrickson, 1979) for ready combination with the phase estimates given by partial structure (Sim, 1959).

Multiple wavelength analysis. It has long been understood that the variation in diffracted intensities near an absorption edge of an anomalous scatterer gives, in principle, a definitive solution for the phase problem. Recent access to synchrotron sources now offers the possibility of meaningful experiments to exploit this effect.

In a phase determination based on multiple wavelength measurements it is essential to relate all data to a common reference point. Karle (1980) showed that when the analysis is referred to the normal scattering terms, the problem takes a rather simple form. We adopt a slight modification of this approach here. Thus, ${}^{\lambda}F(h)$ is a structure factor for reflection h corresponding to the complete factor f at a particular wavelength λ, ${}^{\circ}F(h)$ corresponds only to the normal scattering terms, $f°$, and ${}^{\circ}F_A$ corresponds only to the kth kind of anomalous scatterer in the structure. It is then possible to express the measurable quantities, $|{}^{\lambda}F(h)|^2$, in terms of the separate normal scattering contribution and the atomic scattering factors for each different kind of anomalous scatterer.

It is instructive to examine the case of a single kind of anomalous scatterer. If we denote the phase of ${}^{\circ}F(h)$ by ${}^{\circ}\phi$ and that of ${}^{\circ}F_A(h)$ by ${}^{\circ}\phi_A$, then

$$\begin{aligned}|{}^{\lambda}F(\pm h)|^2 = |{}^{\circ}F(h)|^2 &+ a(\lambda)|{}^{\circ}F_A(h)|^2 \\ &+ b(\lambda)|{}^{\circ}F(h)||{}^{\circ}F_A(h)|\cos({}^{\circ}\phi-{}^{\circ}\phi_A) \\ &\pm c(\lambda)|{}^{\circ}F(h)||{}^{\circ}F_A(h)|\sin({}^{\circ}\phi-{}^{\circ}\phi_A) \qquad (6)\end{aligned}$$

where

$$a(\lambda) = (f^{\Delta}/f^{\circ})^2 \qquad (7a)$$

$$b(\lambda) = 2(\Delta f'/f^{\circ}) \qquad (7b)$$

and

$$c(\lambda) = 2(\Delta f''/f^{\circ}). \qquad (7c)$$

All wavelength dependence in Eq. (6) is embodied in coefficients, Eq. (7), which can be determined in advance of knowledge of the structure. Then from suitable measurements at $\pm$h and at multiple wavelengths, a constrained system of equations (6) can be constructed to solve for $|{}^{\circ}F|$, $|{}^{\circ}F_A|$ and $({}^{\circ}\phi - {}^{\circ}\phi_A)$. The values of $|{}^{\circ}F_A|$ can be used to determine the structure of anomalous scatterers. Calculations from these positions then lead to ${}^{\circ}\phi_A$ and hence to ${}^{\circ}\phi$. In practice, a probabilistic error analysis will also be required.

Direct probabilistic relationships. Until recently, classical direct methods have assumed data for which Friedel's law holds. It now appears that direct relationships are strengthened considerably when anomalous scattering measurements are taken into account. Hauptman (1982) has derived conditional probability distributions for three-phase structure invariants associated with the six distinct structure factors that have indices which obey $h+k+\ell=0$. Eight three-phase invariants are involved and the resulting distributions are unimodal and expected values may occur at any value in the whole interval from 0 to 2π. Giacovazzo (1983) obtained similar results. Karle (1984) has given an alternative analysis in which simple rules are deduced to evaluate triplet phase invariants from particular differences between structure factor magnitudes.

The combination of the phase information from a single-wavelength experiment involving anomalous scatterers with that from direct methods leads to unambiguous estimates of phase invariants. This is also evident from the use of Sayre (1974) refinement in combination with partially resolved anomalous-scattering phase information (Hendrickson and Teeter, 1981). However, the intimate integration of information achieved in the more recent work yields phase information directly without any knowledge about the locations of the anomalous scatterers. This is a distinct difference from the anomalous-scattering methods described in earlier sections. However, in this case it is sums of phases rather than individual phases that are obtained. Thus a global reduction to the phases themselves will be required if the methods are to be used for ab initio phasing.

Tests with the new methods on error-free data indicate that accuracy of estimates can be quite high even for macromolecular structures. This contrasts markedly with convential direct methods.

Computations in Revolved Anomalous Phasing

The efficacy of phase determination by partial-structure resolved anomalous scattering usually depends critically on accurate measurements of relatively small differences between related diffraction intensities. Thus, careful experimental design is of great importance. However, data handling procedures are also very important for optimal phase determination. We have described elsewhere many of the procedures used in our analyses of single-wavelength measurements (Hendrickson and Teeter, 1981; Smith and Hendrickson, 1982; Hendrickson, Smith and Sheriff, 1984). Here the major computational steps are merely summarized.

The first step in the analysis is the reduction of systematic errors by local scaling. This involves a suite of programs including CHKRAT and RMSANO to characterize the distribution of Bijvoet differences as a function of regions of reciprocal space, ANOSCL and SCALE2 to determine and apply parameterized anisotropic local scaling parameters (Hendrickson and Teeter, 1981) and ANORES to evaluate r.m.s. values of ΔF (Eq. 5) as a function of resolution. Comparisons between these values for centric and acentric reflections gives an estimate of signal to noise. The moving window approach to local scaling (Matthews and Czerwinski, 1975) offers an attractive alternative.

The next step is the determination of the structure of anomalous scattering centers. Patterson maps are computed by PATTER from $(\Delta F)^2$ coefficients with appropriate rejection criteria to eliminate outlier data from the calculations. After successful interpreation, the anomalous scattering model is refined by ANOLSQ against ΔF values selected to be good approximations of $|F''_A|$ by virtue of relatively large magnitude. Difference syntheses computed by FORIER are often useful in completing the model.

Once the model is refined, it is possible to prepare the various entities needed for phase determination. It is essential for the probability analysis of partial structure information that the data be on an absolute scale. This scale factor is estimated by Wilson's statistics in KCURVE, which is also used to compute the expected value of structure factor contributions from the unknown part of the structure. SCALE3 applies the absolute scale and flags poorly measured data for special phasing treatment. Next ASCALC calculates the real and imaginary parts of the structure factors for the anomalous scatterers. Then ERRANO finds residual lack-of-closure errors for the anomalous phasing

With everything now in hand, PHASIT is used to determine phases by a probabilistic combination of anomalous-scattering and partial information. This program has the option of multiplicative combination or probabilistic choice in the mixing of the two sources. The program also has the option of producing phases based on either hand of the heavy-atom array (if it is not centric). RESOLV is an alternative phasing program that combines the anomalous scattering with a tentative atomic model of the macromolecule. Calculated phases are then used in FORIER to produce electron density distributions in the two enantiomorphs for interpretation or initiation of further refinement by use of non-crystallographic symmetry, solvent flattening or other density modifications.

Computations in Multiple Wavelength Phasing

Experience in the analysis of measurements at multiple wavelengths about an absorption edge is not as advanced as for the single wavelength case. However, we have recently carried out a successful analysis of synchrotron data at four wavelengths from lamprey hemoglobin. The computations for this problem illustrate the nature of the required analysis.

It is clear from the basic equation, (6), that the scattering factors must be known at the wavelengths of measurement. However, the values very near an absorption edge are not well described by theoretical calculations. Indeed, even the position of the edge depends on the chemical state of the anomalous scatterer. Thus it is essential to evaluate the scattering factors experimentally. This can be done from x-ray absorption spectra measured by fluorescence from the very crystals used for diffraction measurements. A suitable scaling and background correcion must be found to fit these data to the calculated atomic absorption spectrum at points away from the edge. This immediately yields the imaginary part since $\Delta f''$ is proportional to energy times the atomic absorption coefficient. The relationship to produce $\Delta f'$ is more complicated as it involves the

Kramers-Kronig transformation integral. However, with suitable adjustment to theoretical values calculated away from the edge (Cromer, 1983), the scattering factors in the whole range can be evaluated for use in equation (7). The necessary relationships are given in Hendrickson, Smith and Sheriff (1984).

Several possible approaches to the analysis of multiple wavelength data have been proposed. We have chosen in these initial tests to perform a least-squares analysis of the data as described by equation (6) for the case of one kind of anomalous scatterer. Coefficients a, b and c are known and the cofactors of the four terms in (6) are treated as variables subject to the constraint that the structure factor moduli be positive and that the obvious trigonometric identity holds. The process seems to be very robust and yields $|{}^{\circ}F|$, $|{}^{\circ}F_A|$ and $({}^{\circ}\phi - {}^{\circ}\phi_A)$. The values of $|{}^{\circ}F_A|$ can be used to determine the structure of anomalous scatterers by Patterson or direct methods. Then upon calculation of ${}^{\circ}\phi_A$ from these atomic coordinates, ${}^{\circ}\phi$ is known. We have not yet developed a fully appropriate assessment of errors for use in weighting the Fourier coefficients for an electron-density synthesis.

Acknowledgements

I thank Janet Smith, Steven Sheriff, Richard Honzatko and Martha Teeter for their important contributions to the development of the methods desribed here. This work was supported in part by NIH grants GM-29548 and GM-34102.

References

Bijvoet, J. M. (1949). Proc. Acad. Sci. Amst. B52, 313-314.

Cromer, D. T. (1983). J. Appl. Cryst. 16, 437.

Giacovazzo, C. (1983). Acta Cryst. A39, 585-592.

Hauptman, H. (1982). Acta Cryst. A38, 632-641.

Hendrickson, W. A. (1979). Acta Cryst. A35, 245-247.

Hendrickson, W. A., Smith, J. L. & Sheriff, S. (1984). In "Methods in Enzymology: Diffration Methods for Biological Macromolecules", edited by H. W. Wyckoff, C. H. W. Hirs and S. N. Timasheff, in press. Academic Press, New York.

Hendrickson, W. A. & Teeter, M. M. (1981). Nature 290, 107-113.

Karle, J. (1980). Int. J. Quant. Chem. 7, 357-367.

Karle, J. (1984). Acta Cryst. A40, 4-11.

Matthews, B. W. (1966). Acta Cryst. 20, 230-239.

Matthews, B. W. & Czerwinski, E. W. (1974). Acta Cryst. A31, 480-487.

North, A. C. T. (1965). Acta Cryst. 18, 212-216.

Sayre, D. (1974). Acta Cryst. A30, 180-184.

Sim, G. (1959). Acta Cryst. 12, 813-815.

Smith, J. L. & Hendrickson, W. A. (1982). In "Computational Crystallography", edited by D. Sayre, pp. 209-222. Clarendon Press, Oxford.

An Introduction to Electron Density Map Fitting using FRODO.

T. Alwyn Jones(1),&J. Pflugrath(2)

(1)Department of Molecular Biology, Biomedical Centre, Box590, S-75124,Uppsala, Sweden.

(2)Max- Planck- Institute for Biochemistry, D-8033,Martinsried, BRD.

Introduction.

The use of computer graphics as a tool in macromolecular crystallography is now firmly established (Wright, 1982; Jones, 1982; Diamond, 1982). Computer programs have been developed which allow one to build the initial molecular model into an MIR map. Just as importantly, they have been used to make the manual corrections frequently required during subsequent crystallographic refinement. They have rarely been used in making the initial interpretation of how the molecule folds. Our recent success in solving the structure of retinol binding protein (Newcomer et al. 1984) is an exception and most structures are interpreted by first staring at plastic mini-maps until the shape of the molecule becomes clear. This is a pattern recognition problem where one tries to match the varying density shapes and sizes to the amino acid sequence. At low resolution (below 3 A), and often with poor derivatives, this can still be difficult.

In this school, we shall be using the program FRODO (Jones, 1978, 1982, 1984) running on an Evans and Sutherland PS 300 controlled by a Digital Equipment VAX computer. This program is also available on E&S MPS, PS2, Vector General 3400 and MMS-X systems. Our aim is to refit coordinates of Crambin, a small protein whose structure was solved by Hendrickson & Teeter (1981) using the anomalous scattering of the sulphur atoms in three disulphide bridges. Their first model (M1) was built by placing atomic markers on paper plots of their map, proceeding outwards from the disulphide bridges. Not surprisingly, this model has poor stereochemistry, and has an R-factor of 45% with data extending from 7.5 to 1.5 A resolution. After 10 cycles of reciprocal space refinement with PROLSQ (Hendrickson & Konnert, 1980) the new model M11 looks more like a protein and has a reduced R-factor of 35%. However, regions 18-22 and 37-42 still have severe errors that must be manually corrected.

Principle Data Sets.

FRODO is able to display data from three different sources:

1. A molecular data set, which in our case will be coordinates M11. FRODO has a number of ways of picking out portions of this data set and displaying the fragments as lines down between atoms. These lines are drawn if the atoms are closer than a fixed preset value, so that any grouping of atoms can be drawn. Files for M1 and PDB, the final coordinates deposited at the Protein Data Bank, will also be available. If coordinates from M1 are displayed, some chemically sensible bonds will be missing because of poor stereochemistry.

Under VAX/VMS these files are given names such as M11.XYZ.

2. An electron density data set. This type of data set is a file of density values, from which FRODO can pick a volume element to contour at any desired levels. The contouring is made in planes along the three crystallographic axes to give a 3-D mesh affect. This data set is also needed for fingertip refinement of atomic fragments as used by Jones & Liljas (1984) to refine Satellite Tobacco Necrosis Virus.

We will have maps calculated on a 0.8 A grid from models M11 and PDB. They will be calculated with 2 Fo - Fc amplitudes and model phases, and have names such as M11F21.MAP. We will also inspect maps calculated with Fo - Fc as amplitudes, and named M11F11.MAP.

3. Vector data set This could contain groupings of linked vectors representing pre-contoured maps, for example. In this case, the maps could be contoured at a number of levels, any of which could be picked out by FRODO around a specified volume element. Alternatively, it could contain a number of vectorised coordinate data sets any of which can be viewed by specifying the equivalent to a contour level. The data set CRAMBIN.VEC will be available containing vectorised versions of M1, PDB and M11 as levels 1,2 and 3 respectively.

Basic Commands.

There will be a VMS symbol FRODO to start the program. The user will then be prompted for a data set which contains information about such things as other data sets, what the user was doing last, etc. Type FRODO again and wait for a question mark. The user is then in the CHAT interface which responds via 4 character key words, some of which issue prompts. At the first session, you will be set up with M11.XYZ coordinates, M11F21.MAP at a level of 60 units, and the vector data set CRAMBIN.VEC whose access will be switched off. The CHAT commands you will need initially are:

ZONE - to define the start and end residues to be picked out from the coordinate data set. The program prompts for their names. If you want to go to the first very bad place, you should type 17 23 to specify all atoms from residue 17 up to and including 23.

RZON - to define the start and end residues to be regularized so that the chain keeps the correct stereochemistry.

REFI - to activate the regularization option.

ACON - to define the screen centre. It stands for Atom Contour and prompts for the residue and atom name to be used for the screen centre. It will initiate access to the map and vector data sets if previously requested.

DSN2 - to define a different coordinate data set. The name of the file you want to work with must be typed on the next line.

DSN6 - to define a different map data set. The name of the file you want to work with must be typed on the next line.

END - if you regret starting, this will finish the session.

GO - passes control to the display.

The display section of FRODO is also menu driven, but instead of typing commands, one hits them on the screen with a cursor whose position is determined by a pen and tablette. The pen should be held almost vertically, and close to the tablette. Move the pen so that the cursor points to one character in the the command CHAT. Push the pen into contact with the tablette (an alternative action is to sweep the pen over the item to be picked). This activates the command and brings us back to the keyboard in the CHAT interface. Type GO again to return to the screen.

Now move the cursor to a position above an atom and press down. This should identify the atom and its residue provided no other atoms are close by. Practise this for awhile to improve hand to eye coordination.

On the PS300, the view direction, zoom, clipping, picture intensities can be changed by turning the attached dials.

Here are the most important commands:

WAIT - A number of commands could be active (they will have a star next to them). This command clears everything and sets the program back to waiting for something to do.

NAYB - Once activated, identified atoms will generate a fan of dotted lines to their neighbours. It will stay active until turned off.

BOND - Requires two succesful identifications, and then a bond will be drawn between those atoms.

BOBR - bond break. The bond joining the next two identified atoms is broken.

ROCK - picture will rock around a vertical axis. This is an on-off switch, and will stay rocking until it is hit a second time.

The aim of the game is to improve the fit of the coordinates with the density. You can use 4 commands.

MOVE - to move a single atom, the next identified atom. Use the pseudo A-D's or the knobs. The movement X & Y is in screen space, the Z movement is towards or away from you. You will have to change the view a number of times to get the best fit.

FBRT - Foreground/Background/Rotate/Translate - to move a group of atoms. A fragment can be formed using BOND/BOBR. All atoms connected to the identified atom via the present screen connecty, will move as a rigid body. The origin for the rotation operation will be the identified atom. The rotation and translation operations are controlled by the pseudo A-D's and knobs, and are applied in the screen space.

TOR - applies a torsional rotation around the bonds linking atoms. For example, suppose 5 atoms ABCDE are connected A to B to C to D to E. If TOR is activated and then atoms A, B, C, D and E are identified in that order, you should then activate the YES flag to signify you have finished defining the angles. This then allows you to vary the dihedral angles around bonds BC and CD by changing the pseudo A-D's or the knobs. Note that any atoms connected to B other than through the bond BC will not be affected whereas all atoms with connections through to C will move. There must be no closed loops.

The user signifies acceptance or rejection of these 3 commands via the YES/NO flags.

RSR - real space refitting. This allow the computer to move the identified fragment to best fit the density (Jones & Liljas, 1984).

Once a part of the structure has been modified, the user must make a conscious effort to put the results in the molecular data set. This is done with the SAVE command but ensure that neither the YES nor the NO flags are active before pressing SAVE.

Exercise.

Now look at Leu 18 with maps M11F21, M11F11 and then PDBF21. You decide which map to work with (I recommend using PDBF21). Break the bonds 17C-18N and 18C-19N. It should be obvious where the side chain points, so activate FBRT, identify 18CA and fly it in so that the N, CA, CB, C atoms fit nicely, then press YES. Activate TOR, identify 18N, CA, CB, CG and CD1 (or CD2), press YES. Now Chi 1 and 2 can be changed to best fit the density. Atom 18O will also need fitting so use MOVE, or dihedral N, CA, C, O. Once you are happy, press YES and then WAIT to clear all other flags and then SAVE.

Residue 19 is a proline, break 19C-20N bond, FBRT the group, MOVE 19O, SAVE. Repeat for Gly 20 and Gly 21. You will now have quite a nice fitting group of atoms but with incorrect stereochemistry.

This is corrected in FRODO by regularization of the structure to produce ideal bond lengths, angles and dihedral angles. FRODO uses a restrained method (i.e. it never gets the exact values expected in some dictionary) where each atom moves in turn to best fit its local stereochemistry (Hermans & McQueen, 1974). If you wish, use the FIX command on the screen to make sure that an atom will not move out of density on subsequent regularization.

To regularize, enter CHAT, define the refinement zone with RZON, in this case take at least one residue on each side of where you have done the damage, in this case define 17 22. Now type REFI. The program will tell you of its progress as it goes along. At each cycle it informs you of the biggest shift from the starting coordinates and the biggest errors in bond length, angle and dihedral angle. It will prompt twice for listing which may be of interest and then ask if the results should be written into your molecular data set. Answer by typing YES or NO. After returning to CHAT type GO and inspect your handiwork.

It turns out that model M1 was built with the wrong sequence and in particular residue 21 should be a threonine. Did you guess that from the density? Enter CHAT and type SAM. This is another menu driven interface with commands to inspect, modify or copy the coordinate data set. The REPLACE option allows you to mutate the amino acid sequence. The program will prompt you for the name of the residue (in this case 21) and the new residue type (THR). Type END to exit from REPLACE and STOP to exit from SAM.

Unfortunately we are not yet finished with residue 21. In making the REPLACE, the program keeps the information that the residues Gly and Thr have in common, i.e. the atoms N, CA, C, O. To get coordinates for the threonine side chain requires another pass through the regularizer so type REFI. Once you return to the screen you will probably need to rebuild Thr21.

Touch up any poor areas but remember that your aim is to get the coordinates to fit only as well as is required for programs such as PROLSQ and CORELS (Sussman et al., 1977) to make the final changes. In particular, you must fit the peptide planes and the forked side chains correctly with FRODO.

If there is time, move on to region 37 - 42 which may be a bit more difficult.

This tutorial shows some of the things you can do with FRODO. More can be found in the references Jones (1978, 1982, 1984) and the detailed users guide for the graphics type you will be using.

The data sets used in this exercise are available from T.A.Jones. We would like to thank Wayne Hendrickson for making the Crambin data available to us.

Converting FRODO to run on different display types is the result of a collaboration with a number of people, in particular Ian Tickle at Birkbeck College, London on the PS2, Bruce Bush at Merk, New Jersey on the MPS, Jim Pflugrath and Mark Saper at Rice University, Houston on the PS300.

REFERENCES

Diamond, R. (1982) In Computational Crystallography, ed. D. Sayre, pp 318-325, Oxford Univ. Press

Hendrickson, W.A. & Konnert, J.H. (1980) In Computing in Crystallography, eds. R. Diamond, S. Ramaseshan & K. Venkatesan, pp. 1301-1325. Bangalore: Indian Academy of Science

Hendrickson, W.A. & Teeter, M.A. (1981) Nature 290, 107-113

Hermans, J. & McQueen, J.E. (1974) Acta Cryst. A30, 730-739

Jones, T.A. (1978) J. Appl. Cryst. 11, 268-272

Jones, T.A. (1982) In Computational Crystallography, ed. D. Sayre, pp 303-317, Oxford Univ. Press

Jones, T.A. & Liljas, L. (1984) Acta Cryst. A40, 50-57

Newcomer, M.E., Jones, T.A., Aqvist, A., Sundelin, J., Eriksson, U., Rask, L. & Peterson, P. (1984) EMBO Journal 3, 1451-1454

Sussman, J.L., Holbrook, S.R., Church, G.M. & Kim, S.H. (1977) Acta Cryst. A33, 800-804

Wright, W.V. (1982) In Computational Crystallography, ed. D. Sayre, pp 294-302, Oxford Univ. Press

COMBINED CONSTRAINED/RESTRAINED REFINEMENT (CORELS)

A.G.W. Leslie
Blackett Laboratory, Imperial College, London SW7 2BZ, U.K.

1) Introduction

There are several computer programs currently available for the refinement of macromolecules, but only two of these, CORELS (Sussman et al., 1977, Herzberg & Sussman, 1983) and the Gauss-Seidel program of Hoard and Nordman (1979), have the facility of being able to refine the parameters of constrained groups of atoms rather than refining individual atomic parameters. Within a constrained group, all bond lengths and angles are fixed at values determined from small-molecule crystallography. In CORELS the refined parameters are the the position and orientation of the groups and, optionally, defined dihedral angles within the groups. (In this sense, a constrained group is not necessarily rigid.) Similarly, temperature factors are applied to groups of atoms rather than individual atoms. This approach results in a considerable reduction in the number of refined parameters, and, as a consequence, a significant improvement in the data to parameter ratio. This has the effect of increasing both the speed and the radius of convergence.

In its original form, as applied to the refinement of a tRNA molecule, CORELS was limited to the refinement of rigid groups against the X-ray data alone. Although the refinement was partially successful, the refined models invariably displayed poor stereochemistry at the junctions between constrained groups and many unfavourably short contacts between non-bonded atoms. To overcome this difficulty, harmonic spring-like restraints were introduced to regularise the stereochemistry at the junctions between constrained groups and to minimise the number of overshort interatomic contacts. In addition, the sparse matrix conjugate gradient procedure was implemented to reduce the amount of computation required. The latest version of the program also features keyworded input, which makes the program very much easier to use.

The function minimised in the refinement is

$$Q = \omega_F DF + \omega_D DD + \omega_T DT \qquad (1)$$

where the first term is the usual structure factor contribution:

$$DF = \sum_h \omega_h \left(|F_{obs,h}| - |F_{calc,h}|\right)^2$$

The second term is for the stereochemical restraints, which are all expressed as distance restraints between pairs of atoms:

$$DD = \sum_{d} \omega_d \, (D_{obs,d} - D_{calc,d})^2$$

where $D_{obs,d}$ is the "ideal" distance between two atoms and $D_{calc,d}$ is the distance calculated from the model. (Planarity restraints are dealt with in this way by the introduction of dummy atoms.)

The final term restrains the model to a set of target coordinates, and is expressed as:

$$DT = \sum_{i} \omega_i \, (|\underline{X}_{targ,i} - \underline{X}_{calc,i}|)^2$$

$\underline{X}_{targ,i}$ is a vector defining the position of the $\underline{i}$th atom in the target coordinates and $\underline{X}_{calc,i}$ is the vector defining its position in the model.

The values assigned to the weights $\omega_F, \omega_D, \omega_T$ in (1) depend on the type of refinement being undertaken. For model building, ω_F is set to zero while for structure refinement ω_T will normally be zero.

The quantity Q in equation (1) is an explicit function of all the group positional and thermal parameters, and can be written as:

$$Q = f(t, R, \psi, B)$$

In this expression t and R refer to the translation vector and rotation matrix to be applied to each constrained group, ψ describes the internal dihedral angles of each group and B represents the thermal parameters. Group derivatives are obtained by differentiation of (1) with respect to individual parameters and application of the chain rule, as described in Sussman et al., 1977.

The choice of constrained groups will depend on the resolution of the X-ray data included in the refinement. At very low resolution (6 to 8Å), an entire domain may be treated as a rigid group, while at higher resolution (3.5 to 4Å) individual elements of secondary structure such as α helices or strands of β sheet can be refined. At still higher resolution (above 3.5Å), the constrained groups will usually correspond to individual amino-acid residues, and the internal dihedral angles will be additional parameters in the refinement.

The choice of the initial resolution range for the X-ray data will depend on the origin of the starting model. If it has been obtained by building into a m.i.r. phased electron density map then refinement can be initiated using data to 3.5Å or 3Å resolution. If, on the other hand, the model has been obtained from a related structure (ie a different structural form of the same enzyme or the structure of a similar enzyme from another source), then it is advisable to start refinement with a few rigid bodies, possibly corresponding to domains, and low resolution data (6 to 8Å). In either case the resolution of the X-ray data can be extended as the refinement proceeds.

2) Applications of CORELS

A survey of the literature reveals that although CORELS refinement has been applied to a wide range of protein structures in several different laboratories, it has primarily been used to position an entire molecule, treated as a single rigid group, within the unit cell. Further structure refinement has almost invariably been performed using one of the restrained least squares refinement programs (Hendrickson and Konnert, 1980, Jack and Levitt, 1978), with alternate rounds of refinement and manual rebuilding on an interactive graphics display.

Although experience has shown that such an approach is, in general, successful, it is not obvious that this procedure is necessarily the most economical route to a final high-resolution refined structure.

In this paper I would like to describe two rather different examples of the way in which CORELS has been used to considerable advantage in protein structure refinement.

The first example, which will be described in rather greater detail, represents a rather special class of structural problem. There are many well documented examples of proteins which undergo a significant conformational change on ligand binding (eg hexokinase, liver alcohol dehydrogenase, citrate synthase, glyceraldehyde 3-phosphate dehydrogenase (GAPDH)). In these cases the conformational change can often be approximated by a rigid body rotation of one domain with respect to another. If a well refined structure of the enzyme in one state of ligation is available, then CORELS can be used to obtain a refined structure of the enzyme in other state(s) of ligation. This application will be illustrated by recent work on the structure determination of the apo and 1 NAD per tetramer forms of GAPDH using the holo-enzyme structure as a starting model (Leslie & Wonacott, 1984).

The second example is the way in which CORELS is used routinely by the Uppsala group to eliminate errors in a structural model before going on to use a restrained least squares refinement program.

3) The application to GAPDH

a) Introduction

GAPDH is a tetramer of four chemically identical subunits, with a total molecular weight of 145000, which requires the cofactor nicotinamide adenine dinucleotide (NAD) for activity. The structure of the enzyme from the thermophile Bacillus stearothermophilus with four molecules of NAD bound to the tetramer (the holo-enzyme) has been determined at 2.7Å resolution (Biesecker et al., 1977). The four subunits of the tetramer are arranged with 222 molecular symmetry, and each subunit consists of two domains (Fig.1): the coenzyme binding domain, which binds the NAD molecule, and the catalytic domain, which contains all the residues involved in catalysis. The catalytic domains form the core of the tetramer. The holo-enzyme has recently been refined at 2.4Å resolution to a crystallographic residual of 22% using the Hendrickson-Konnert refinement program (unpublished work of A.J.Wonacott and P.C.E. Moody). This structure provided an initial model for the structure determination of the apo-enzyme and the enzyme with one molecule of NAD bound to the tetramer.

A 6Å resolution isomorphous replacement map of the apo-enzyme showed that a significant conformational change occurs on the loss of bound NAD (Wonacott and Biesecker, 1977). The conformational change was interpreted as a rigid body rotation of the coenzyme binding domain of each subunit with respect to the catalytic domain, with a movement of up to 5Å in the helices α_D and α_E. The apparent rigid body nature of the movement suggested that the problem would be well suited to refinement using CORELS.

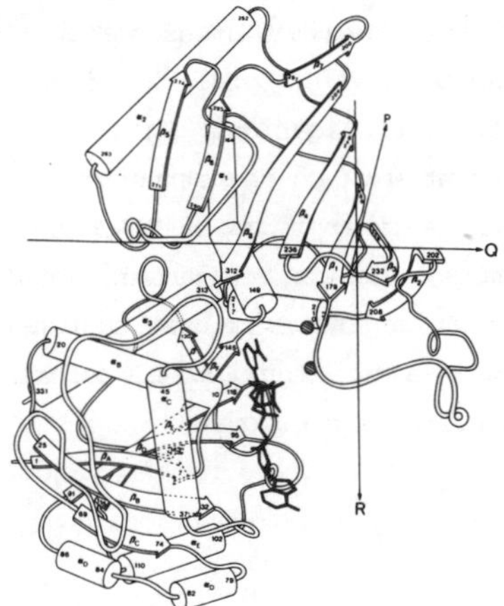

Figure 1.
A schematic representation of the structure of one subunit of B. stearothermophilus holo-GAPDH

b) Initial CORELS refinement of the apo-enzyme

The positions of the derivative heavy atom sites were used to locate the origin and orientation of the the molecular 222 symmetry axes in the monoclinic apo unit cell. (In principle this could have been done equally well using the rotation and translation functions.) This information was used to position the refined holo-enzyme structure in the unit cell as a starting model for refinement. Initially X-ray data between 10Å and 6Å resolution were included (3096 reflections) and each subunit was divided into three rigid groups corresponding to the coenzyme-binding domain excluding residues 312-333, the catalytic domain (residues 148-311) and the C-terminal helix (residues 312-333). The crystallographic asymmetric unit contains the entire tetramer, giving a total of 72 positional parameters for refinement, corresponding to a data to parameter ratio of about 40:1. In spite of this high ratio the magnitudes of the parameter shifts were rather different for the same rigid groups in different subunits, suggesting that the apo-enzyme does not possess 222 molecular symmetry. This did not seem wholly reasonable, particularly since the positions of the derivative heavy atom sites displayed excellent 222 symmetry. It seemed more probable that the apparent asymmetry was artefactual, arising from deficiences in the model and the X-ray data. A procedure was therefore developed to impose 222 molecular symmetry constraints on the refinement (Leslie, 1984).

c) Application of molecular symmetry constraints to the CORELS refinement

There are several different ways in which molecular symmetry information can be incorporated into a least-squares refinement procedure. For example, the linked-atom least-squares program LALS (Smith and Arnott, 1978), which is used to refine polymeric molecules using fibre diffraction data, imposes helical symmetry by Lagrangian constraints. The Hendrickson-Konnert program applies molecular symmetry restraints by restraining each subunit to an average structure. This approach, although very general, has the disadvantage that the form of the restraints inevitably restrains atomic parameters to their current values. In a CORELS refinement, where shifts of several angstroms may be required, this could seriously hinder the rate of convergence. If, however, the molecular symmetry is imposed by a suitable averaging of the parameter shifts calculated by CORELS then this difficulty is avoided. This is the basis of the method adopted, and is best illustrated by a simple two-dimensional example.

In Fig. 2, two objects A and B whose positions are to be refined are assumed to be related by a two-fold axis passing through the origin and normal to the plane of the paper. The positional and rotational shifts to be applied to object A are denoted $\delta X_A, \delta Y_A, \delta\phi_A$ and similarly for object B. If the refined positions of A and B are to obey the two-fold symmetry then we require:

$$\begin{aligned} \delta X_A &= -\delta X_B \\ \delta Y_A &= -\delta Y_B \\ \delta\phi_A &= \delta\phi_B \end{aligned} \qquad (2)$$

If the calculated shifts $\delta X_A{}^c$, $\delta Y_A{}^c$, $\delta\phi_A{}^c$ etc do not satisfy the conditions (2) , it is possible to impose the two-fold symmetry on the refined positions of A and B by averaging the calculated shifts and setting:

$$\begin{aligned} \delta X_A &= -\delta X_B = 0.5(\delta X_A{}^c - \delta X_B{}^c) \\ \delta Y_A &= -\delta Y_B = 0.5(\delta Y_A{}^c - \delta Y_B{}^c) \\ \delta\phi_A &= \delta\phi_B = 0.5(\delta\phi_A{}^c + \delta\phi_B{}^c) \end{aligned}$$

The same type of argument can easily be extended to three dimensions. In order to apply these constraints to a CORELS rigid-group refinement the following steps are necessary.

i) The parameters explicitly refined by CORELS for each rigid group are three rotation angles which define a rotation matrix $[R]$ and a vector $\underline{t}$ denoting the position of the centre of gravity of the group. The vector $\underline{t}$ is expressed in fractional crystallographic coordinates, while the matrix $[R]$ is applied to angstrom coordinates in an orthogonal frame based on the crystallographic axes. In order to apply the shift-averaging procedure required to impose molecular symmetry, it is first necessary to find the rotation matrix $[S]$ and translation vector $\underline{T}$ which produce an equivalent shift when applied to coordinates expressed in an orthogonal frame based on the molecular symmetry axes rather than the crystallographic axes. In the case of 222 molecular symmetry, this frame is defined by the three two-fold symmetry axes. This transformation ensures that there is a simple

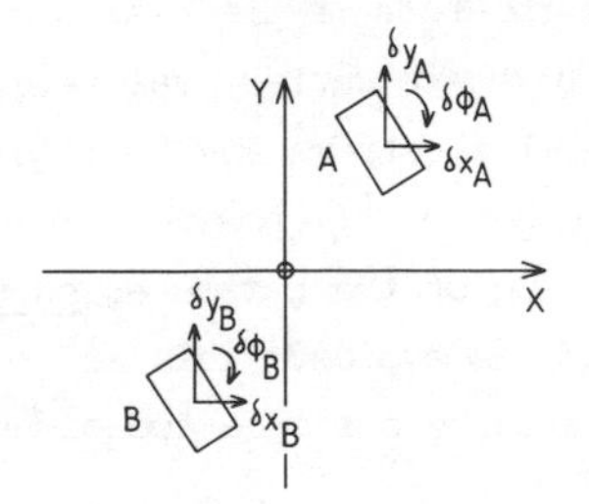

Figure 2.

relationship between the parameter shifts (given by $[S]$ and $\underline{T}$) for rigid groups related by molecular symmetry (eg of the kind in equations (2)).
ii) One of the subunits is chosen as a reference subunit. For each rigid group in the reference subunit, the matrix $[S]$ and vector $\underline{T}$ for the <u>same</u> rigid group in all other subunits are transformed, so that the transformed parameters $[S_R]$ and $\underline{T}_R$ would produce the structurally equivalent shift when applied to the coordinates of the group in the reference subunit. If the molecular symmetry operation relating subunit $\underline{i}$ to the reference subunit is denoted $[M_i]$, then the so-called <u>reduced</u> matrix and vector are defined as:

$$[S_R] = [M_i]^{-1} [S] [M_i]$$

$$\underline{T}_R = [M_i]^{-1} \underline{T}$$

iii) If the refined molecule is to possess the molecular symmetry of the starting model, then $[S_R]$ and $\underline{T}_R$ must be identical for the same rigid group in all subunits. In practice this will not be the case, and so the molecular symmetry is imposed by forcing all the $[S_R]$ and $\underline{T}_R$ to be identical. This is simply achieved by making the reduced vectors for all subunits equal to the simple average of the individual vectors. The rotation matrices $[S_R]$ can be decomposed into the three rotation angles, and then these angles are similarly averaged over all the subunits.
iv) The equivalenced parameters $[S_R]$ and $\underline{T}_R$ are transformed back first to give $[S]$ and $\underline{T}$ and then back to the original CORELS parameters $[R]$ and $\underline{t}$. When these "averaged" $[R]$ and $\underline{t}$ are applied to the starting coordinates they will provide a set of refined coordinates which exactly obey the molecular symmetry.

d) <u>The effect of molecular symmetry constraints on the 6Å refinement of GAPDH</u>

To test the effectiveness of the molecular symmetry constraints the apo-enzyme model with three rigid groups per subunit was refined at 6Å resolution for nine cycles both with and without the molecular symmetry constraints. The results are presented in Fig.3. It is clear that the constrained refinement is more stable and converges more rapidly then the unconstrained refinement. This is particularly true for the C-terminal helix (plots $\underline{c}$ and $\underline{d}$). Furthermore, a comparison of the alpha carbon coordinates of the two refined models with the final 4Å refined structure indicated that the constrained model was closer to the true structure.

e) The final 4Å resolution refinement of the apo-enzyme

The 6Å resolution refinement reduced the R factor from its initial value of 44.5% down to 39.4%, at which point it became clear that the structure would have to be divided into a larger number of rigid groups to achieve a further improvement. A number of different ways of dividing up the structure were attempted, leading to the final set of rigid groups shown in Table 1, which was used to refine the structure at 4Å resolution. In retrospect it is clear that it would have been possible to go directly from the 6Å, three groups per subunit refinement to the 4Å, ten groups per subunit refinement without any intervening stages, and this course was followed successfully for the 1 NAD per tetramer enzyme. The final R factor for the 222 symmetric model was 29.7%. The molecular symmetry constraints were then released and two further cycles reduced the R factor to 28.7%. The resulting asymmetry is localised primarily in those regions of the molecule that are involved in intermolecular contacts in the crystal lattice.

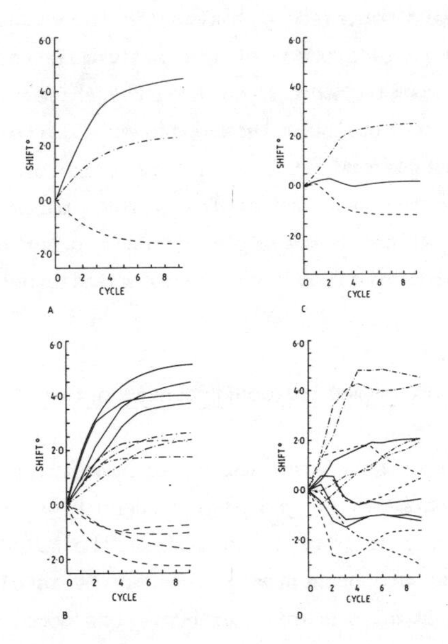

Figure 3. The variation of the refined rotational parameters ϕ(dashed), θ(full) and ρ(dash-dot) for the coenzyme binding domain (a and b) and the C-terminal helix (c and d). (a) and (c) represent results of the constrained refinement, (b) and (d) the unconstrained refinement, both at 6Å resolution. The four lines for each parameter in (b) and (d) correspond to the four subunits.

Table 1. The rigid groups used in the 4Å CORELS refinement of GAPDH.

CORELS group	Secondary Structural units	Residue numbers
1	β_A,β_B,β_C	0-8,23-34,71-77
2	α_B	9-22
3	α_C	36-52
4	β-structure	53-70
5	α_D	78-88
6	α_E	101-113
7	β_D,β_E,β_F	89-100,114-120,142-147
8	$\beta_E-\beta_F$ loop	121-141
9	catalytic domain	148-311
10	C-terminal helix	312-333

Stereochemical restraints were included but only with very low weight. They were used as indicators of distortions in the stereochemistry rather than as effective restraints on the refined model. The r.m.s. deviation in peptide bond length between rigid groups was 0.49Å (mean 0.28Å) with a maximum value of 1.6Å for the bond between β_B and α_C. Only an overall temperature factor was refined.

The atomic coordinates were combined with individual atomic temperature factors from the holo-enzyme model, and structure factors calculated to 3Å resolution (R factor 30.2%). A difference electron density map was calculated at 3Å resolution and averaged over the four subunits. This map clearly indicated deficiences in the model and the positions of forty water molecules per subunit. This model is suitable for high-resolution refinement using a restrained least squares program.

f) CORELS refinement of the 1 NAD per tetramer enzyme

The apo-enzyme was used as a starting model for the refinement of the 1 NAD enzyme which crystallises in an orthorhombic space group. Derivative heavy atom positions were again used to position the molecule in the unit cell, and refinement was initiated using data to 6Å resolution and three rigid groups per subunit. Eight cycles of refinement reduced the R factor from 40.3% to 37.1%. A further eight cycles at 4Å resolution using ten rigid groups per subunit (Table 1) reduced the R factor from 38.5% to 29.5%. The resulting coordinates were combined with individual atomic temperature factors from the holo-enzyme, and the resulting model was refined for eleven cycles using the Hendrickson-Konnert program and data to 3Å resolution. This refinement eliminated all the distortions in the stereochemistry introduced by the CORELS refinement and reduced the R factor to 23.9% for all data to 3Å. A minor rebuild is now required.

4) The application of CORELS to the refinement of other protein structures

CORELS has been used routinely by the group at Uppsala to refine models obtained by model building into m.i.r. phased electron density maps. The technique has been applied successfully to two crystal forms of liver alcohol dehydrogenase, the C-terminal fragment of L7/L12 ribosomal protein, carbonic anhydrases B and C, serum retinol binding protein and phage T4 thioredoxin.

The strategy employed is to use CORELS to eliminate as many of the errors in the model as possible before going on to use a restrained least squares refinement program. In the initial stages of refinement X-ray data to at least 3Å resolution are included, and the constrained groups are individual amino acid residues. Stereochemical restraints are included at a sufficiently high weight to avoid serious distortions, but because the geometry within each residue is constrained errors in the model become localised in the peptide bonds linking constrained groups. To give sufficient flexibility to the model an r.m.s. deviation in peptide bond lengths of up to 0.1Å is permitted, rather than the usual figure of 0.025Å for a restrained refinement. Deviations in planarity and the peptide bond angle are correspondingly greater. Only two temperature factors per residue are applied, one for main chain and one for side chain atoms. Typically this gives a data to parameter ratio of 1.7:1 at 3Å resolution or 3:1 at 2.5Å resolution. Cycles of CORELS refinement are alternated with manual rebuilding on a graphics display system. As the refinement proceeds, data to higher resolution are included (if available) and the process continued until there are no further interpretable features on $2F_o-F_c$ maps. This usually requires a total of about 5 to 10 cycles of refinement with rebuilding after every second cycle.

Further refinement with a restrained least squares program usually gives a further reduction of 3-4% in the R factor. Most of this improvement results from changes in thermal parameters and small movements (0.1-0.2Å) of atoms in the peptide bonds. In addition the stereochemical errors in the model which are localised in the peptide bonds in the CORELS refinement become more evenly distributed.

5) Conclusions

The application of CORELS refinement to GAPDH has produced models for the enzyme in two other states of ligation by NAD. These models are suitable for further refinement using a restrained least squares program. The use of CORELS has made it possible to solve both structures without the need for high-resolution isomorphous derivative data and with an absolute minimum of time-consuming manual model building. These features make this approach particularly attractive for this type of structural problem.

In more general applications the results from the Uppsala group suggest that CORELS can also be used in the early stages of refinement of any protein structure. The very rapid convergence of the CORELS program results in a more economical use of computing resources than refinement with a restrained least squares program. However, no refinement program will remove large non-random errors in coordinates; these must still be dealt with by manual correction (Jones, 1982)

Acknowledgements

I would like to thank Joel Sussman for providing the latest key-worded version of CORELS and Alwyn Jones for communicating the refinement results of the Uppsala group.

References

BIESECKER,G.,HARRIS,J.I.,THIERRY,J.C.,WALKER,J.E. & WONACOTT,A.J. (1977) Nature (London),266,328-333.

HENDRICKSON,W.A. & KONNERT,J.H. (1980). Computing in Crystallography, edited by R. Diamond, S. Ramaseshan & K. Venkatesan, pp 13.01-13.23. Bangalore: Indian Academy of Sciences.

HERZBERG,O. & SUSSMAN,J.L. (1983). J. Appl. Cryst. 16, 144-150.

HOARD,L.G. & NORDMAN,C.E. (1979). Acta Cryst. A35, 1010-1015.

JACK,A. & LEVITT,M. (1978). Acta Cryst. A34, 931-935.

JONES,T.A. (1982). Computational Crystallography, edited by D. Sayre, pp 303-317. Clarendon Press, Oxford.

LESLIE,A.G.W. (1984). Acta Cryst. A40 in press.

LESLIE,A.G.W. & WONACOTT,A.J. (1984). J. Mol. Biol. in press.

SMITH,P.J.C. & ARNOTT,S.A. (1978). Acta Cryst. A34, 3-11.

SUSSMAN,J.L.,HOLBROOK,S.R.,CHURCH,G.M. & KIM,S.H. (1977). Acta Cryst. A33, 800-804.

RESTRAINED LEAST-SQUARES REFINEMENT OF PROTEINS

Wayne A. Hendrickson

Department of Biochemistry and Molecular Biophysics
Columbia University, New York, NY 10032, U.S.A.

and

Laboratory for the Structure of Matter
Naval Research Laboratory, Washington, D. C. 20375, U.S.A.

Careful refinement of the crystal structures of proteins and other macromolecules has come to be recognized in recent years as a crucially important step in the crystallographic analysis of such molecules. The resulting precision in detail is essential to the understanding of functional properties such as reaction mechanisms and to the design of specifically altered molecules for synthesis by site directed mutagenesis. However, refinement also usually proves to be an essential ingredient in the structure determination itself as initial interpretations are generally wrong in some respects.

The theory that relates the geometric parameters of an atomic model to its crystallographic diffraction pattern is very sound. This makes it possible in principle (and also in practice for favorable cases) to refine an atomic model to a match of high fidelity with the experimental diffraction data. While rigorous structure refinement has long been a routine procedure in small-molecule crystallography, several factors frustrate the straightforward extension to crystals of macromolecules. First, the sheer size of the computational problem is daunting. Second, most macromolecular crystals diffract relatively weakly. Hence, typically there is a paucity of observable data and this severaly restricts the degree of overdetermination in the problem. Third, initial models of macromolecular structures are usually quite inaccurate. This poor start coupled with the non-linear character of the equations generally results in convergence to a false minimum.

The incorporation of prior knowledge about the sterochemistry of biological macromolecules helps greatly to overcome the special difficulties of macromolecular refinement. This knowledge can supplement the limited diffraction data and thereby make the refinement problems better conditioned. Good geometry also facilitates the manual rebuilding by computer graphics that is needed to escape false minima.

Several different methods have been used to include stereochemistry in the refinement process. The various approaches have been reviewed elsewhere (Hendrickson and Konnert, 1980). At present the most commonly used refinement programs are ones described initially by Jack and Levitt (1978), by Sussman et al. (1977) and by us (Konnert, 1976; Hendrickson and Konnert, 1980). The purpose of this article is to describe some of the computational aspects of the method of stereochemically restrained least-squares refinement of protein molecules as we have developed the method. Many details of these procedures have been described previously (Konnert, 1976; Hendrickson and Konnert, 1980; Konnert and Hendrickson, 1980; Hendrickson and Konnert, 1981; Hendrickson, 1981; Wlodower and Hendrickson, 1982; Hendrickson, 1984). Thus only the rudiments of the method will be outlined here. In addition, a few of the special computational techniques used in the programs will be described.

Fundamentals of the Method

The method of restrained least-squares refinement treats knowledge about the local geometry in a macromolecular atomic model as observations with expected values derived from crystallography of the component structures, from spectroscopy or possibly from theoretical considerations. Each of these observations is assigned a variance that estimates the presumed breadth of the distribution about the expected value for the feature. The stereochemical observations are included together with the diffraction data in a composite observational function,

$$\phi = \phi_{diffraction} + \phi_{bonding} + \phi_{planarity} + \cdots , \tag{1}$$

which has terms such as

$$\phi_{diffraction} = \sum_{refns} \frac{1}{\sigma_F^2} (|F_{obs}| - |F_{calc}|)^2 \tag{2}$$

and

$$\phi_{bonding} = \sum_{dists} \frac{1}{\sigma_D^2} (d_{ideal} - d_{model})^2 . \tag{3}$$

Weighting is given by the inverse of variances, σ^2.

The observational function is to be minimized with respect to the variable parameters -- in particular the atomic coordinates, thermal parameters and occupancy factors for some atoms. Inasmuch as most of the observational terms are non-linear, the minimization is done iteratively by Taylor series linearization to yield normal equations

$$A \, \delta = b \tag{4}$$

that involve changes, δ, in the parameters. Contributions from the different terms in (1) are simply additive in the normal matrix, A, and right-hand-side vector, b. Thus,

$$(A_{\text{diffraction}} + A_{\text{bonding}} + \cdots\cdot) \, \delta = (b_{\text{diffraction}} + b_{\text{bonding}} + \cdots). \tag{5}$$

The stereochemical terms are observations of a rather special kind. They link the structural parameters together much more directly than do the structure factor equations. Accordingly we refer to these observations as restraints that effectively reduce the degree of freedom available to the model. The structure of the method leads to a very facile inclusion of additional classes of restraints. Presently, we include restraints related to bonding distances, planarity of groups, chirality at asymmetric centers, non-bonded contacts, restricted torsion angles, non-crystallographic symmetry and thermal parameters. Many of these yield observational functions that are equivalent to terms in typical potential energy descriptions. The detailed formulation of these restraints is given in earlier publications.

Program Design

Our implementation of restrained refinement is really a system of programs rather than a single entity. There are two main programs, PROLSQ and PROTIN as well as several ancillary service routines and some special purpose programs.

PROLSQ (PRotein Least SQuares) is the actual refinement program. It reads diffraction data and scatterng factors prepared by SCATT (SCATTering data), initial atomic coordinates and restraint specifications prepared by PROTIN (PROTein model INput), parameter shifts from previous refinement cycles, and control card-images. It then augments the normal-equation elements pertinent to each of the stereochemical restraints and the structure factor observations. Fractional atomic coordinates are used in order to speed the rate-limiting calculations concerning structure factors. For the same reason, a highly optimized space-group specific routine, CALC, is used for computing structure factors and their derivatives. Elements of the resulting sparse normal-matrix are stored in a singly dimensioned array that is indexed by pointers. Next, PROLSQ uses a conjugate-gradients procedure to solve the new parameter shifts. Finally, it tests the expected impact of the new shifts on the R-value. An optimal shift damping factor is searched for in trials against a sample of the data. Alternative versions of PROLSQ use FFT procedures to compute the diffraction contributions. Mitchell Lewis has adapted routines from Jack and Levitt (1978) whereas Barry Finzel has used ones from Agarwal (1978).

PROTIN is run once before a series of refinement cycles to prepare the atomic coordinate data needed by PROLSQ and to identify the atoms and ideal values involved in the individual stereochemical restraints. It incidentally also peforms a useful verification function for initial models. Ideal values for the various stereochemical features are taken from those in particular small-molecule crystal structures of constituent parts of the macromolecule.

The system also includes STANDARD, a program to generate new ideal groups for PROTIN from crystal data for amino acids; CONEXN, a program to generate the various restraint specifications for a prosthetic group; HAFFIX, a program to affix hydrogen atoms on the heavy-atom skelton of a protein and ROTLSQ, a program to perform a rigid-body least-squares refinement of a model.

Computational Techniques

Effective implementation of the restrained least squares method has involved several special computational techniques. It seems appropriate to mention a few of these here. One important aspect is the use of group dictionaries for the specification of ideal stereochemical parameters. This is patterned after the early model-building program by Diamond (1966). A set of standard groups has been assembled from well-refined neutron structures of the components, and the source structures are listed by Wlodawer and Hendrickson (1982). Linkage groups such as peptides and disulfide bridges are specified in a similar manner. Atomic identifications and target ideal values for such features as bonding distances and chiral volumes is implicit in these structures, and appropriate tables are deduced in PROTIN for the complete set of such features. These dictionaries are then consulted to produce the specifications for a particular polymeric structure. This is a convenient and versatile procedure, but suffers the disadvantage of being based on specific structures rather than on average values. This could cause an unwanted bias in the final stages of refinement. However, it would be possible to produce Euclidean components that conform optimally to a set of average geometrical parameters. Variances used for weighting parameters can also be deduced from the averaging process. For example, the distribution of chiral volumes at α-carbons of amino acids is found to have a standard deviation of 0.15Å^3.

Another important feature is the handling of non-bonded contact identification. For large structures the check of all pairwise interactions can become very time-consuming. Here, a table of possible contacts (all non-bonded pairs within some distance, e.g. 4.5Å) is generated in PROTIN and only these are used in PROLSQ. The work in PROTIN is made expeditious by first determining the centroid and maximal radius of each amino-acid residue. Then each residue pair is checked for possible contact and then, but only if any exist, the individual atomic pairs are checked. This is a structurally logical analog of the "cubing" algorithm sometimes used for such searches.

The impact of the size of macromolecules is especially acute in the dimensions of the full normal matrix. Since a typical protein structure may have a few thousand atoms, the full normal matrix would have several million elements. However, the structure of the restrained least-squares problem is such that only those elements associated with bonded or contact distance have appreciable value. These are only a fraction of a percent of the total. The conjugate gradients procedure for solution of the system of linear equations (4) nicely takes full advantage of this sparseness. However, a special indexing scheme is required for economy in storage. We use a single-dimensional array with a series of pointers to specify the location of particular elements. In addition, it is necessary to maintain a pointer system to identify the location of individual distances since, for example, the same interatomic pair might arise as a bond, in a plane, in a chiral center and in a torsion angle.

It is worth noting that although PROLSQ is quite general, PROTIN is written specifically for proteins. The extension to other macromolecules is rather straightforward and Gary Quigley has written a corresponding NUCLIN for nucleic-acid refinement. However, it is clearly important to revise PROTIN to accommodate either kind of polymer since protein:nucleic-acid complexes represent an important class of structures under study.

References

Agarwal, R. C. (1978). Acta Cryst. A34, 791-809.

Diamond, R. (1966). Acta Cryst. 21, 253-266.

Hendrickson, W. A. (1981). In "Refinement of Protein Structures", edited by P. A. Machin, J. W. Campbell and M. Elder, pp. 1-8. Daresbury Laboratory, Warrington.

Hendrickson, W. A. (1984). In "Methods in Enzymology: Diffraction Methods for Biological Macromolecules", edited by H. W. Wyckoff, C. H. W. Hirs and S. N. Timasheff, in press. Academic Press, New York.

Hendrickson, W. A. & Konnert, J. H. (1980). In "Computing in Crystallography", edited by R. Diamond, S. Rameseshan and K. Venkatesan, pp. 13.01-13.23. Indian Academy of Sciences, Bangalore.

Hendrickson, W. A. & Konnert, J. H. (1981). In "Biomolecular Structure, Function and Evolution", edited by R. Srinivasan, Vol. 1, pp. 43-57. Pergamon, Oxford.

Jack, A. & Levitt, M. (1978). Acta Cryst. A34, 931-935.

Konnert, J. H. (1976). Acta Cryst. A32, 614-617.

Konnert, J. H. & Hendrickson, W. A. (1980). Acta Cryst. A36, 344-350.

Sussman, J. L., Holbrook, S. R., Church, G. M. & Kim, S.-H. (1977). Acta Cryst. A33, 800-804.

Wlodawer, A. & Hendrickson, W. A. (1982). Acta Cryst. A38, 239-247.

Index